Mohamed Elzagheid
Macromolecular Chemistry

Also of Interest

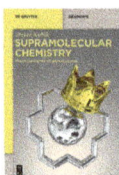

Supramolecular Chemistry.
From Concepts to Applications
Stefan Kubik, 2020
ISBN 978-3-11-059560-4, e-ISBN 978-3-11-059561-1,
e-ISBN (EPUB) 978-3-11-059357-0

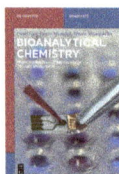

Bioanalytical Chemistry.
From Biomolecular Recognition to Nanobiosensing
Paolo Ugo, Pietro Marafini, Marta Meneghello, 2021
ISBN 978-3-11-058909-2, e-ISBN 978-3-11-058916-0,
e-ISBN (EPUB) 978-3-11-058929-0

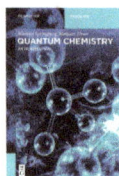

Quantum Chemistry.
An Introduction
Michael Springborg, Meijuan Zhou, 2021
ISBN 978-3-11-074219-0, e-ISBN 978-3-11-074220-6,
e-ISBN (EPUB) 978-3-11-074223-7

Green Chemistry.
Principles and Designing of Green Synthesis
Syed Kazim Moosvi, Waseem Gulzar Naqash, Mohd. Hanief Najar,
2021
ISBN 978-3-11-075188-8, e-ISBN 978-3-11-075189-5,
e-ISBN (EPUB) 978-3-11-075203-8

Host-Guest Chemistry
Brian D. Wagner, 2020
ISBN 978-3-11-056436-5, e-ISBN 978-3-11-056438-9,
e-ISBN (EPUB) 978-3-11-056439-6

Mohamed Elzagheid

Macromolecular Chemistry

Natural and Synthetic Polymers

DE GRUYTER

Author
Prof. Dr. Mohamed Elzagheid
Chemical and Process
Engineering Technology
Jubail Industrial College
Jubail Industrial City
Jubail 31961
Saudi Arabia
elzagheid_m@jic.edu.sa

ISBN 978-3-11-076275-4
e-ISBN (PDF) 978-3-11-076276-1
e-ISBN (EPUB) 978-3-11-076279-2

Library of Congress Control Number: 2021944635

Bibliographic information published by the Deutsche Nationalbibliothek
The Deutsche Nationalbibliothek lists this publication in the Deutsche Nationalbibliografie;
detailed bibliographic data are available on the Internet at http://dnb.dnb.de.

© 2022 Walter de Gruyter GmbH, Berlin/Boston
Cover image: Mike_Kiev / iStock / Getty Images Plus
Typesetting: Integra Software Services Pvt. Ltd.
Printing and binding: CPI books GmbH, Leck

www.degruyter.com

Preface

This book is vastly based on lectures I gave at the Faculty of Pharmacy, University of Benghazi "Garyounis University," and Chemical and Process Engineering Technology Department, Jubail Industrial College.

When writing this textbook, the following two initial goals were set:

- To have a book that combines both natural and synthetic polymers and at the same time fulfills the needs of many students at the graduate and advanced undergraduate levels in the fields of organic chemistry, bioorganic chemistry, polymer sciences, industrial chemistry, pharmacy, and biological sciences.
- To design a book in a way that even students who have little or no background in macromolecular chemistry can still gain knowledge they need and proceed further with their studies.

Macromolecular chemistry topics in this book are presented at the level required for organic chemistry, biochemistry, and polymer chemistry majors. The material of this book was also discussed in a simple way to facilitate the understanding of the discussed topics by other students who study in related or different majors.

It is advisable that when students begin to browse through macromolecular chemistry book while seeing a large number of organic compounds, names, and reactions, not to try to understand everything at once, they need to do it step by step or topic by topic and remember the following golden rules:

- Attending the class is crucial.
- Asking the teacher about questions that come to mind or if something is unclear is very important.
- Reviewing book chapters on a weekly basis, if possible, or chapter by chapter and try having discussions with classmates on the topics already studied will for sure enrich the knowledge and give a better understanding of the topics discussed in each chapter.

I hope instructors will find this book a good reference for their teaching and also hope that students get a good experience in learning different topics from different chapters of this book.

Mohamed Ibrahim Elzagheid
Waterloo, Canada
July 2021

https://doi.org/10.1515/9783110762761-202

Contents

Part II: Synthetic Polymers

Part I: **Natural Polymers**

Chapter 1
Introduction

1.1 Macromolecules

A macromolecule is a large molecule (polymer) with a large molecular mass. The units are usually bonded or connected together by covalent bonds such as O-glycosidic (found in carbohydrates), N-glycosidic (found in nucleic acids), and peptide bonds (found in polypeptides and proteins) (Figure 1.1).

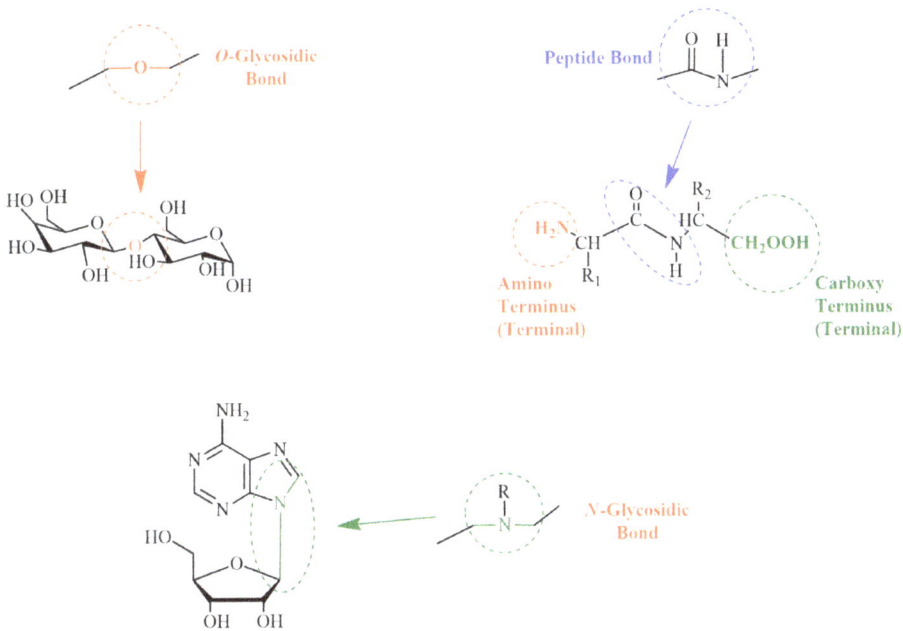

Figure 1.1: Examples of macromolecules and how they are connected by O-glycosidic, N-glycosidic, and peptide bonds.

In general, the use of macromolecule term covers biopolymers such as proteins, polysaccharides, nucleic acids, lipids, and industrial polymers such as polyethylene oxide (PEO), polyethylene (PE), polypropylene (PP), and other synthetic polymers. Biopolymers are natural polymers that are found in nature, and some of their analogs can be made in the laboratory. Industrial polymers are man-made polymers that are synthesized from small organic and inorganic molecules, and most of them are made from oil and gas products. Examples of these polymers are shown in Figure 1.2.

https://doi.org/10.1515/9783110762761-001

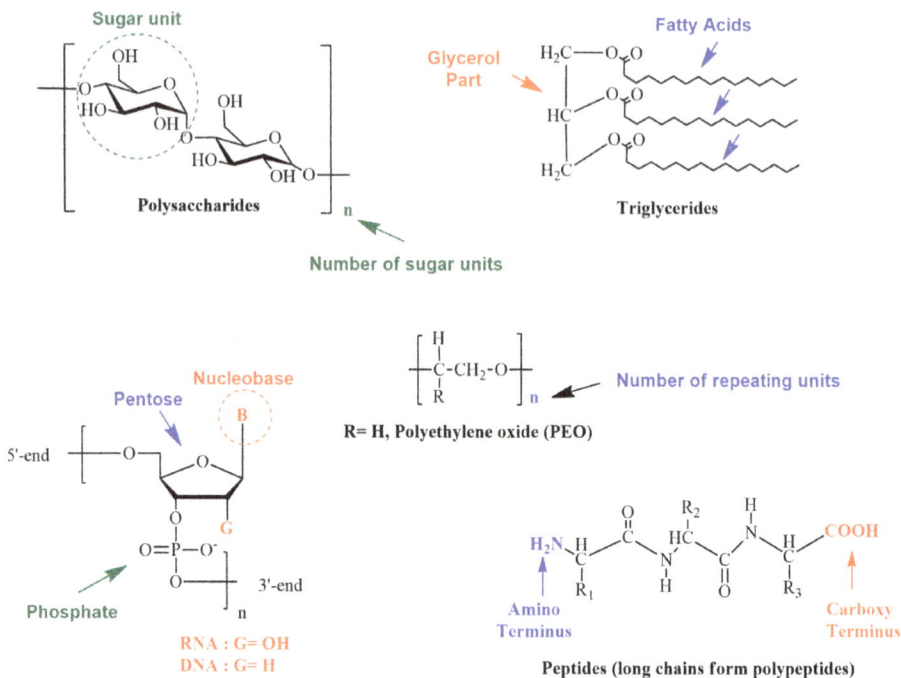

Figure 1.2: Examples of biopolymers and synthetic polymers.

1.2 Biopolymers (Natural) and Synthetic (Industrial) Polymers

1.2.1 Examples of biopolymers are carbohydrates (sugars), proteins (peptides and polypeptides), and nucleic acids (deoxyribonucleic acid (DNA) and ribonucleic acid (RNA)); and the monomer units are monosaccharides, amino acids, and nucleotides, respectively. Lipids (fats and oils) have triglycerides as building units as shown in Table 1.1.

Table 1.1: Examples of selected polymers and their monomeric units.

Polymers Macromolecules	Monomeric units Building blocks	Atoms present in the structure
Carbohydrates (sugars)	Mono/disaccharides	C, H, O
Proteins	Amino acids	C, H, O, N
Nucleic acids (DNA and RNA)	Nucleotides	C, H, O, N, P
Lipids (fats and oils)	Triglycerides	C, H, O

1.2.1.1 Carbohydrates can either have simple structures as in monosaccharides or complex structures as in disaccharides, oligosaccharides, and polysaccharides. Their size can range from four carbons as in tetrose to six carbons as in hexoses. They can also exist as aldoses or ketoses based on the functional groups they have in addition to polyhydroxy groups. Carbohydrates can be reducing or nonreducing sugars. They can undergo different reactions such as oxidation and reduction.

1.2.1.2 Lipids are a diverse group of organic compounds. They are insoluble in water but dissolve in nonpolar solvents such as ether, hexane, and acetone. Most lipids have fatty acids in their structure except steroids which are made of tetracyclic carbon rings. Fatty acids are long-chain carboxylic acids, and they can be saturated or unsaturated based on the presence or absence of double bonds in the structure. Triglycerides are formed when glycerol is esterified by three fatty acids. Lipids can be classified into waxes, triglycerides (triacylglycerols), phospholipids (glycerophospholipids), sphingolipids (glycolipids and sphingomyelins), eicosanoids (prostaglandins and leukotrienes), and steroids.

1.2.1.3 Proteins are the polymers of amino acids. In the structure, each amino acid has two functional groups, namely, amine and a carboxyl group. Based on the side chain, amino acids can have different names and different physical and chemical properties. Amino acids can be polar, nonpolar, acidic, or basic. The names of amino acids can be either abbreviated by a three-letter code or one-letter code. For example, the alanine name can be abbreviated as Ala or A. All amino acids are chiral except glycine because it has only three different groups attached to the central carbon. They can form dipeptides by the condensation of two amino acids and polypeptides by the condensation of more amino acids. Polypeptides are usually made in the lab through solid-phase peptide synthesis.

1.2.1.4 Nucleic acids are the polymers of nucleotides. There are two types of nucleic acids, namely, RNA and DNA. Each nucleotide is made up of three parts: nucleobase, pentose sugar, and phosphate group. There are five types of nucleobases: uracil, thymine, cytosine, guanine, and adenine. Pentose sugars are deoxyribose and ribose. There are different types of modified nucleic acids. Among those are the peptide (or peptido) nucleic acid (PNA), morpholinos nucleic acid (MNA) glycol/glycerol nucleic acids, threose nucleic acid, and 4′-thionucleic acids (4′-SDNA). Nucleic acids have many applications and among those are the polymerase chain reaction and DNA fingerprinting.

1.2.2 Industrial or synthetic polymers are made from small materials. Polymers can be classified based on their origin, structure, mode of polymerization, and molecular forces. They are made by polymerization, which converts monomers to polymer. The types of polymerizations are condensation and addition. There are three common condensation polymers, namely, polyesters, polyamides (PA), and polycarbonates (PC). Addition polymerizations can be classified further into free radical polymerization, ionic polymerization (cationic and anionic), and coordination polymerization. The addition polymerization reaction consists of three steps: initiation, propagation, and termination. There are four polymerization techniques such as bulk, solution, suspension, and emulsion. Examples of synthetic polymers available in the industry are polyethylene terephthalate (PET), polyvinyl chloride, polylactic acid, PA, PE, PEO, polyacrylic acid (PAA), PP, polypropylene oxide (PPO), polyvinyl acetate, PC, and polystyrene.

1.3 Chemical Bonding

The two main ways in which atoms can be combined to form molecules are electrovalent bonding (ionic bonds) or covalent bonding (covalent bonds). Some molecules contain both electrovalent and covalent bonds, but many have just one or the other type. There are also much weaker attractions between atoms in molecules called hydrogen bonds. Hydrogen bonding is usually found in nucleic acids and proteins.

1.3.1 Bond Classification Based on Difference in Electronegativity

Bonds are classified on the basis of their electronegativities into covalent, ionic, and polar covalent as shown in Table 1.2.

Table 1.2: Bonds classification.

Bond type	Difference in electronegativity	Examples
Covalent	0	Cl–Cl (3.16–3.16 = 0) H–H (2.20–2.20 = 0)
Ionic	≥2	NaCl (3.16–0.93 = 2.23) KBr (2.96–0.82 = 2.14)
Polar covalent	0< and <2 (greater than 0 and lesser than 2)	H–S–H (2.58–2.20 = 0.38) H–O–H (3.44–2.20 = 1.24)

1.3.2 Electrovalent Bond

Electrovalent (ionic) bonds are formed by an electrical attraction between positively charged cations and negatively charged anions as shown in Figure 1.3.

Sodium Metal
Na
Neutral

Na
Ion

Cl
Ion

NaCl
Oppositely charged ions are hold together to form sodium chloride

Chlorine Gas
Cl-Cl
Neutral

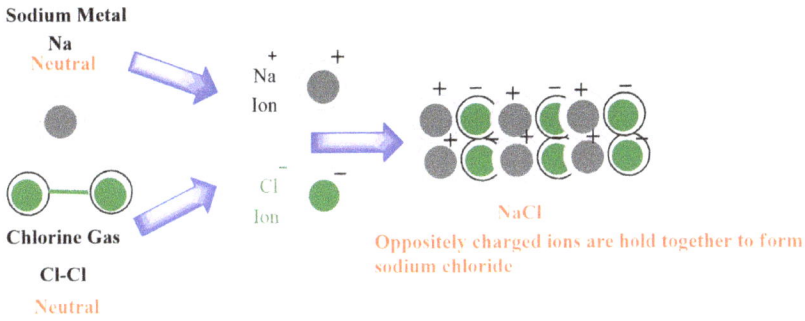

Figure 1.3: NaCl ionic bond formation.

1.3.3 Covalent Bond

Covalent bonds are formed by sharing electrons between atoms. An example is the methane gas, CH_4, shown in Figure 1.4.

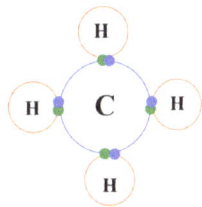

H

H   C   H

H

● Electron from carbon
● Electron from hydrogen   **Figure 1.4:** Covalent bond electron sharing in methane.

1.3.4 Hydrogen Bond

Hydrogen bonds are formed by attractions between the positively charged hydrogen atom with either oxygen or nitrogen atom as shown in Figure 1.5.

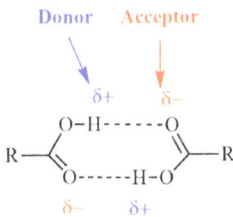

Donor   Acceptor

$\delta+$ $\delta-$
O—H------O
R—   —R
O------H—O
$\delta-$ $\delta+$

Two molecules of carboxylic acids

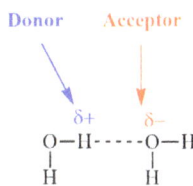

Donor   Acceptor

$\delta+$ $\delta-$
O—H-----O—H
H   H

Two molecules of water

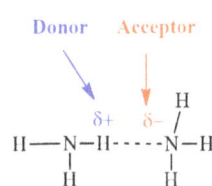

Donor   Acceptor

H
$\delta+$ $\delta-$
H—N—H-----N—H
H   H

Two molecules of ammonia

Figure 1.5: Hydrogen bond formation between different molecules.

1.4 Functional Groups

Some functional groups can play an important role in the structure and function of the macromolecules (polymers). Examples of these groups are listed in Figure 1.6.

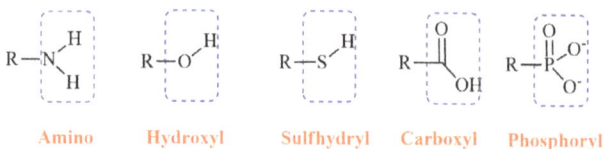

Figure 1.6: Examples of different functional groups seen in macromolecules.

1.5 Types of Chemical Reactions in Macromolecules

1.5.1 Glycoside Bond Formation (Glycosidation Reaction)

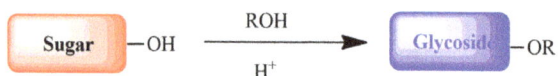

1.5.2 Triglyceride Formation (Esterification Reaction)

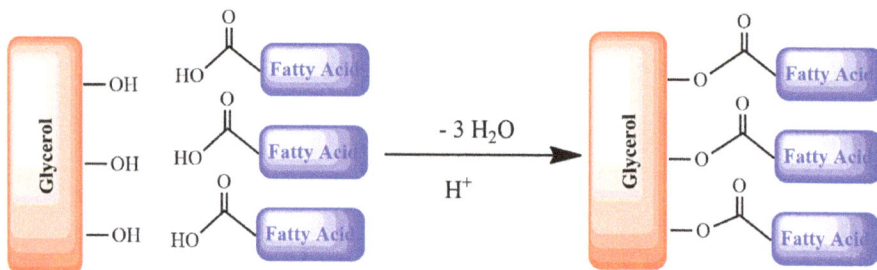

1.5.3 Peptide Bond Formation (Condensation Reaction)

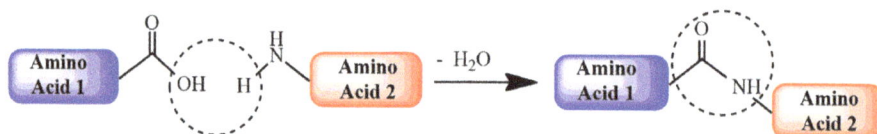

1.5.4 Dinucleotide or Oligonucleotide Formation (Coupling Reaction)

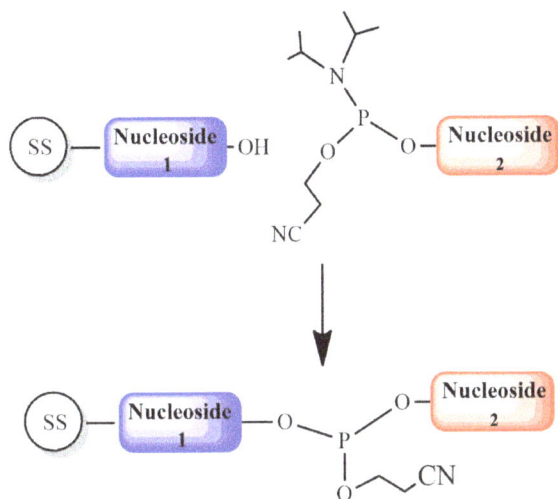

1.5.5 Polymer Formation (Polymerization Reaction)

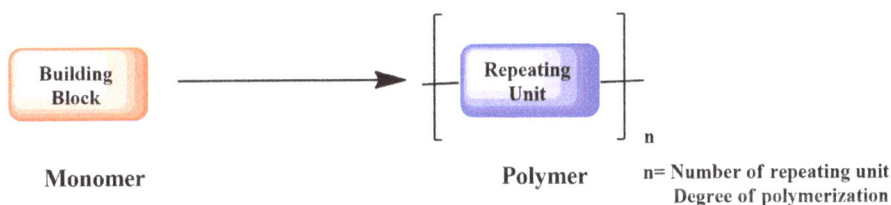

1.6 Drawing Chemical Structures of Macromolecules

Drawing the chemical structures of macromolecules is not an easy task because they have different building blocks, and each building block is made up of multiple functional groups and sometimes a few structural units. So, the easy way to have a complete structure is to build up separate single units and then link them together by linkages. One of the best tools for drawing chemical structures is the **ChemDraw** software which can be easily used to draw different structures of macromolecules. There are different chemical structural formulas that are used to present different structures such as the Kekulé, line (skeletal), and condensed.

The straightforward and time-saving one is the skeletal formula. For example, drawing the chemical structure of vitamin **A** using both Kekulé and skeletal formulas shows clearly which one best fit and takes less time (Figure 1.7).

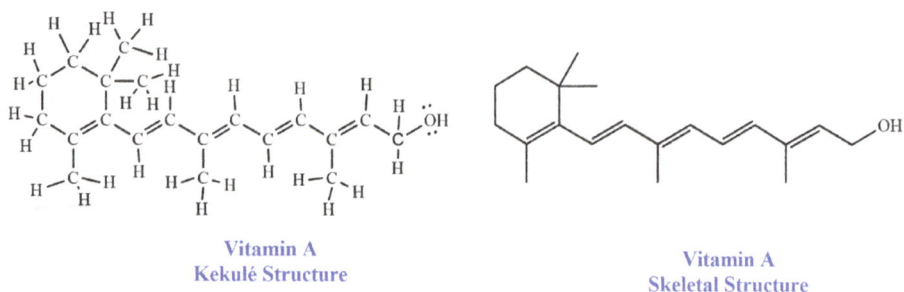

Vitamin A
Kekulé Structure

Vitamin A
Skeletal Structure

Figure 1.7: Kekulé and skeletal structures of vitamin A.

1.7 Summary and Important Concepts

Macromolecule. This is a large molecule (polymer) with a large molecular mass made from small monomers.

Biopolymers. These are natural molecules that include carbohydrates, proteins (peptides and polypeptides), nucleic acids (DNA and RNA), and lipids.

Carbohydrates. They are polymers of monosaccharides and can form complex structures of disaccharides, oligosaccharides, and polysaccharides.

Lipids. These are a diverse group of organic compounds made from triglycerides which in turn formed from the esterification of glycerol with three fatty acids. Some lipids do not have fatty acids in their structure.

Proteins. These are the polymers of amino acids and can exist as small peptides or large polypeptides.

Nucleic acids. These are the polymers of nucleotides. There are two types of nucleic acids: RNA and DNA.

Industrial polymers. These are large molecules made from small synthetic materials.

Chemical bonding. There are two main ways by which atoms can be combined to form molecules. These are electrovalent bonding to form ionic bonds and covalent bonding to form covalent bonds.

1.8 Practice Exercises

1.8.1 Briefly describe the covalent bond and give one example.

1.8.2 Complete the missing information in the following table:

Biopolymers	Monomers
Carbohydrates	
	Amino acids
Nucleic acids	

1.8.3 Write down the full names of the following polymers:
i.   PET
ii. PPO
iii. PAA

1.8.4 Write down the abbreviations for the following polymer names:

Acronym	Name
	Polyethylene
	Polypropylene
	Polystyrene
	Polyvinyl chloride
	Polycarbonate

1.8.5 Briefly explain how a sodium ion (Na^+) of a sodium chloride (NaCl) becomes hydrated or solvated in water.

1.8.6 Show how hydrogen bonds are formed between two molecules of ammonia.

1.8.7 What type of molecule is released during the peptide bond formation?

1.8.8 Name the following functional groups that play an important role in the structure and function of the biomolecules:

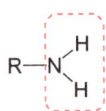

1.8.9 Show the hydrogen bond donor and hydrogen bond acceptor in the following hydrogen bonding scheme between two molecules of ammonia.

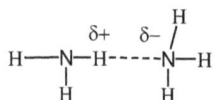

1.8.10 Show the O-glycosidic, N-glycosidic, and peptide bonds in the following structures:

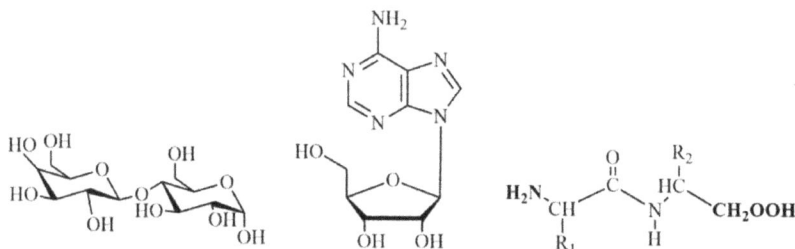

1.8.11 Explain how metal salts are dissolved in water.

1.8.12 How many carbon atoms are there in the following sugars?
i.   Pentose
ii. Tetrose
iii. Triose

1.8.13 Define the hydrogen bond. Give two examples for molecules that can form hydrogen bonds.

1.8.14 Explain why it is better to use skeletal formula when drawing the structures of macromolecules and give one example different from the one given in the chapter above.

1.8.15 What are the differences in the electronegativities in the following molecules?
i.   NaCl
ii. Cl_2
iii. H_2S

1.8.16 How industrial polymers are classified?

1.8.17 List down the two types of polymerization.

1.8.18 What are the classes of lipids?

1.8.19 What are the two types of natural nucleic acids?

1.8.20 Write down the full names of the following artificial nucleic acids:
i.   MNA
ii. PNA
iii. SDNA

1.8.21 Show how hydrogen bonds are formed between two molecules of carboxylic acids and show the hydrogen bond donor and the hydrogen bond acceptor.

1.8.22 What are the functional groups involved in the peptide bond formation?

1.8.23 Are the repeating unit and building block structure similar in the polymerization reaction?

1.8.24 Why are three fatty acids required for triglyceride formation?

1.8.25 What are the main atoms present in the nucleic acid structures?

Chapter 2
Carbohydrates

2.1 Definition

Carbohydrates are compounds containing C, H, and O. Their general formula is $C_x(H_2O)_y$ or $C_n(H_2O)_m$. All have C=O and OH functional groups.

Sugars, or carbohydrates, are characterized by having aldehyde (CHO) or ketone (C=O) groups in addition to two or more hydroxyl (OH) groups. The simplest carbohydrates are glyceraldehydes and dihydroxyacetones as shown in Figure 2.1.

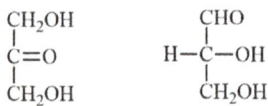

$$
\begin{array}{cc}
\text{CH}_2\text{OH} & \text{CHO} \\
| & | \\
\text{C}{=}\text{O} & \text{H}{-}\text{C}{-}\text{OH} \\
| & | \\
\text{CH}_2\text{OH} & \text{CH}_2\text{OH} \\
\\
\text{Dihydroxyacetone} & \text{Glyceraldehyde}
\end{array}
$$

Figure 2.1: Chemical structures of glyceraldehydes and dihydroxyacetones.

2.2 Classification

2.2.1 On the Basis of the Number of Sugar Units

Carbohydrates can be classified based on the number of sugar units as follows:
- Monosaccharides = single sugar unit

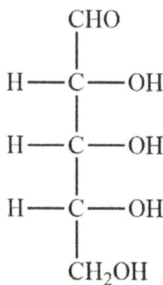

$$
\begin{array}{c}
\text{CHO} \\
| \\
\text{H}{-}\text{C}{-}\text{OH} \\
| \\
\text{H}{-}\text{C}{-}\text{OH} \\
| \\
\text{H}{-}\text{C}{-}\text{OH} \\
| \\
\text{CH}_2\text{OH}
\end{array}
$$

- Disaccharides = two sugar units

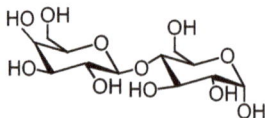

https://doi.org/10.1515/9783110762761-002

- Oligosaccharides = 3–10 sugar units

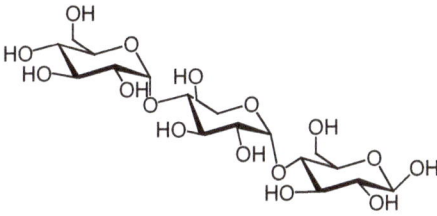

- Polysaccharides = more than 10 units

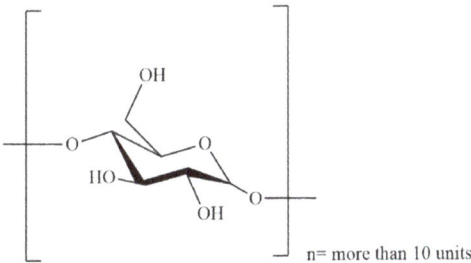

n= more than 10 units

2.2.2 Further Classification of Monosaccharides

2.2.2.1 Based on the counting of the number of carbon atoms present in the structure, monosaccharides can be classified into trioses with three carbons, tetroses
with four carbons, pentoses with five carbons, and hexoses with six carbons as
shown in Figure 2.2.

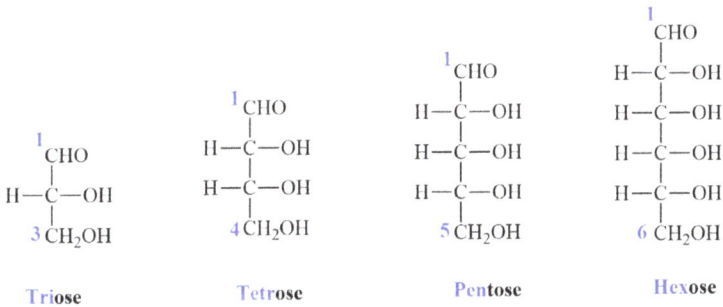

Figure 2.2: Chemical structures of triose, tetrose, pentose, and hexose.

2.2.2.2 Based on the type of carbonyl group that exists in the structure, carbohydrates can be classified into aldoses or ketoses. An aldose has an aldehyde functional group and ketose has a ketone functional group as shown in Figure 2.3.

Aldehyde group

$$\begin{array}{c} H \\ | \\ C=O \\ | \\ H-C-OH \\ | \\ H-C-OH \\ | \\ H-C-OH \\ | \\ CH_2OH \end{array}$$

Aldose (Aldo-sugar)

$$\begin{array}{c} CH_2OH \\ | \\ C=O \quad \text{Ketone group} \\ | \\ HO-C-H \\ | \\ H-C-OH \\ | \\ H-C-OH \\ | \\ CH_2OH \end{array}$$

Ketose (Keto-sugar)

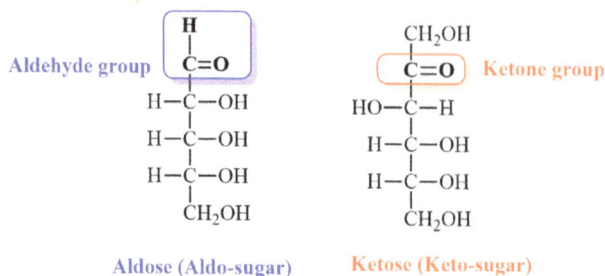

Figure 2.3: Chemical structures of aldose and ketose.

2.2.2.3 D and **L** designations of sugars are done based on their relationships to glyceraldehyde structure. The carbon chain is numbered from the carbonyl group end of the molecule, and the highest numbered chiral center is used to determine **D** and **L** configurations as presented in Figure 2.4.

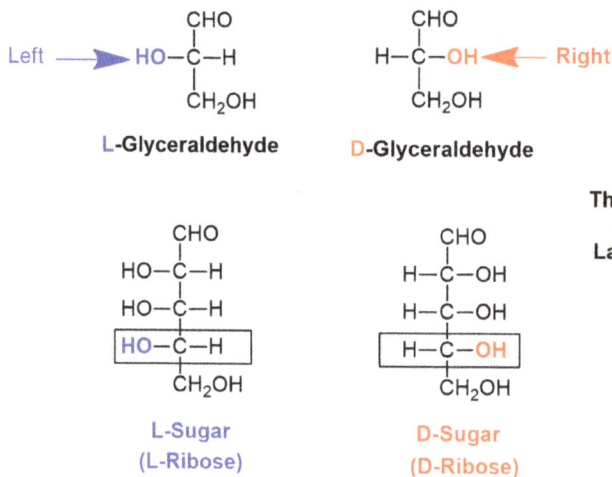

Left ⟶

$$\begin{array}{c} CHO \\ | \\ HO-C-H \\ | \\ CH_2OH \end{array}$$

L-Glyceraldehyde

$$\begin{array}{c} CHO \\ | \\ H-C-OH \\ | \\ CH_2OH \end{array}$$ ⟵ Right

D-Glyceraldehyde

$$\begin{array}{c} CHO \\ | \\ HO-C-H \\ | \\ HO-C-H \\ | \\ \boxed{HO-C-H} \\ | \\ CH_2OH \end{array}$$

L-Sugar
(L-Ribose)

$$\begin{array}{c} CHO \\ | \\ H-C-OH \\ | \\ H-C-OH \\ | \\ \boxed{H-C-OH} \\ | \\ CH_2OH \end{array}$$

D-Sugar
(D-Ribose)

The D- and L- system is named after the Latin Dexter and Laevus, which translate to left and right.

Figure 2.4: D and L configurations.

2.3 Special Carbohydrates

2.3.1 Monosaccharides

2.3.1.1 D-Glucose
- A monosaccharide (aldohexose sugar).
- Its common names are dextrose, grape sugar, and blood sugar.
- This is the most important sugar in our diet and the most abundant organic compound found in nature.

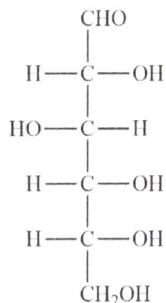

$$
\begin{array}{c}
\text{CHO} \\
|\\
\text{H}-\text{C}-\text{OH} \\
|\\
\text{HO}-\text{C}-\text{H} \\
|\\
\text{H}-\text{C}-\text{OH} \\
|\\
\text{H}-\text{C}-\text{OH} \\
|\\
\text{CH}_2\text{OH}
\end{array}
$$

D-Glucose

2.3.1.2 D-Galactose
- A monosaccharide (aldohexose sugar).
- It is also called brain sugar because it is a component of some chemicals that exist in brain and nerve tissues.
- It is also present in the chemical markers that distinguish various types of blood – A, B, AB, and O.

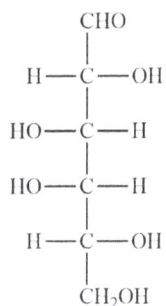

$$
\begin{array}{c}
\text{CHO} \\
|\\
\text{H}-\text{C}-\text{OH} \\
|\\
\text{HO}-\text{C}-\text{H} \\
|\\
\text{HO}-\text{C}-\text{H} \\
|\\
\text{H}-\text{C}-\text{OH} \\
|\\
\text{CH}_2\text{OH}
\end{array}
$$

D-Galactose

2.3.1.3 D-Fructose
- Monosaccharide (ketohexose sugar).
- Its common names are fruit sugar and laevulose (levulose).
- It is the sweetest of all sugars and found in honey and many fruits (e.g., blackberry and raspberry).

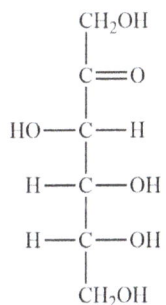

$$
\begin{array}{c}
\text{CH}_2\text{OH} \\
|\\
\text{C}=\text{O} \\
|\\
\text{HO}-\text{C}-\text{H} \\
|\\
\text{H}-\text{C}-\text{OH} \\
|\\
\text{H}-\text{C}-\text{OH} \\
|\\
\text{CH}_2\text{OH}
\end{array}
$$

D-Fructose

2.3.1.1 Structural Formulae of Monosaccharides

Fischer projections are used to represent stereochemistry in open-chain monosacchar-ides. In those projections, the horizontal lines are understood to project out of the plane toward the reader, and vertical lines are understood to project behind the plane. There are also circle-and-line formula and wedge-line-dashed wedge formula (Figure 2.5).

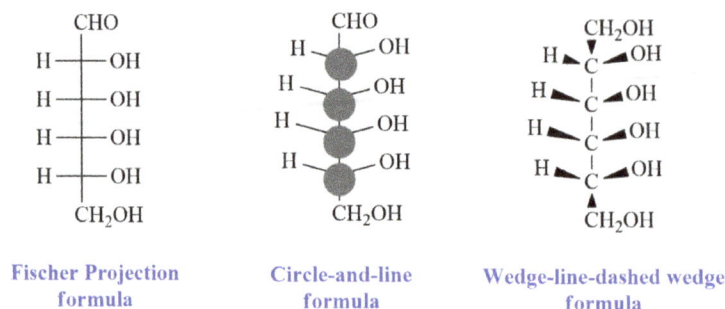

Fischer Projection formula   Circle-and-line formula   Wedge-line-dashed wedge formula

Figure 2.5: Fischer projection, circle-and-line, and wedge-line-dashed wedge formulas.

On the other hand, cyclic monosaccharides are represented by the Haworth formula and chair conformation shown in Figure 2.6.

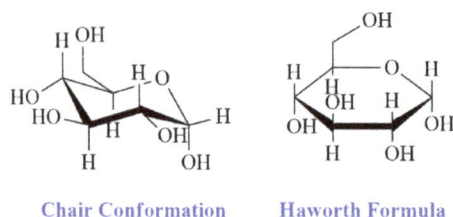

Chair Conformation   Haworth Formula

Figure 2.6: Haworth formula and chair conformation.

2.3.1.2 Cyclic Forms of Monosaccharides

The preferred structural forms of many monosaccharides are the cyclic hemiacetals. There are two types of cyclic structures, that is, furanose (five-membered) and pyra-nose (six-membered), reflecting the ring size relationship to the common heterocy-clic compounds furan and pyran as shown in Figure 2.7.

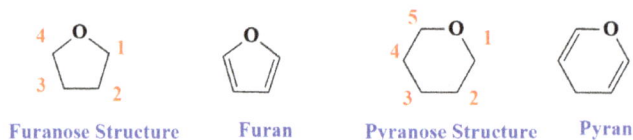

Furanose Structure   Furan   Pyranose Structure   Pyran

Figure 2.7: Cyclic forms of monosaccharides.

A good example is the five-membered cyclic hemiacetal β-D-ribofuranose.

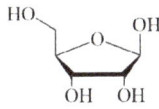

Another example is the six-membered cyclic hemiacetal β-D-glucopyranose.

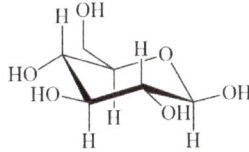

2.3.1.3 Cyclization of D-Glucose

The cyclization proceeds in two steps. In step 1, the linear aldehyde of D-glucose is tipped on its side, and rotation about the C4–C5 bond brings the C5-hydroxyl function close to the aldehyde carbon. While in step 2, the attack by OH group attached to C-5 on the C=O will lead to the formation of β- and β-D-glucopyranose (Figure 2.8).

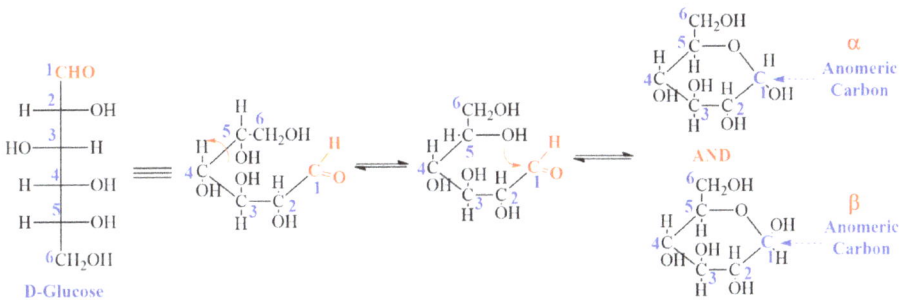

Figure 2.8: Cyclization of D-glucose.

2.3.1.4 Cyclization of D-Fructose

The cyclization proceeds in two steps. In step 1, the linear aldehyde of D-fructose is tipped on its side, and rotation about the C4–C5 bond brings the C5-hydroxyl function close to the aldehyde carbon. While in step 2, the attack by OH group attached to C-5 on C=O will lead to the formation of α- and β-D-fructofuranose (Figure 2.9).

2.3.1.5 Cyclization of D-ribose

The cyclization proceeds in two steps. In step 1, the linear aldehyde of the D-ribose is tipped on its side, and rotation about the C3–C4 bond brings the C4-hydroxyl function close to the aldehyde carbon. While in step 2, the attack by OH group attached to C-4 on C=O will lead to the formation of α- and β-D-ribofuranose (Figure 2.10).

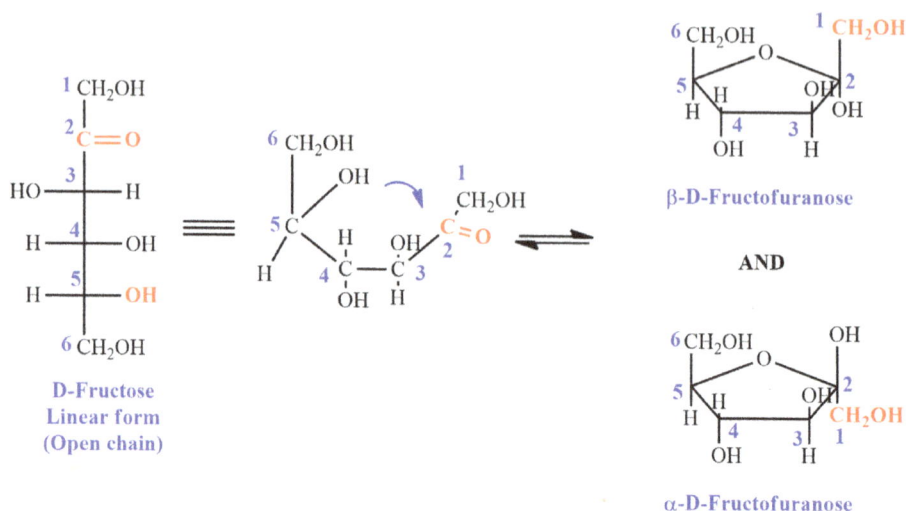

Figure 2.9: Cyclization of D-fructose.

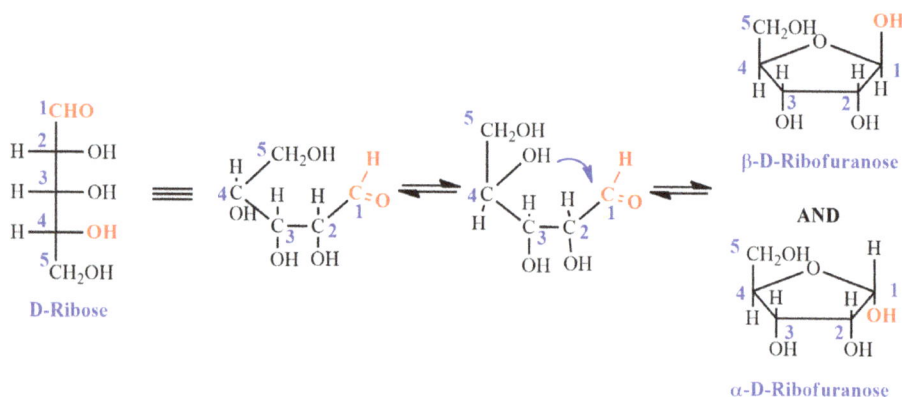

Figure 2.10: Cyclization of D-ribose.

2.3.1.6 Reactions of Monosaccharides

2.3.1.6.1 Glycosidation (glycoside bond formation)

In chemistry, a glycosidic bond is a certain type of functional group that joins a carbohydrate (sugar) molecule to an alcohol, which may be another carbohydrate. Specifically, a glycosidic bond is formed between the hemiacetal group of a saccharide (or a molecule derived from a saccharide) and the hydroxyl group of some alcohol. Glycosides are acetals at the anomeric carbon of carbohydrates. When glucose reacts with an alcohol (e.g., methanol) in the presence of catalytic acid, the methyl glycoside is obtained. A glycoside made from glucose is called a glucoside as shown in Figure 2.11.

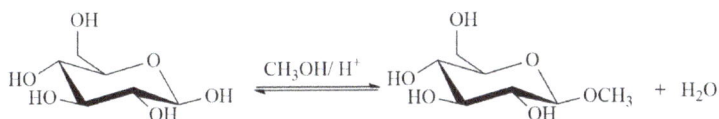

Figure 2.11: Methyl glucoside formation.

2.3.1.6.2 Hydrolysis of glycosides

When glycosides are subjected to acidic conditions, hydrolysis takes place leading to a mixture of anomers as shown in Figure 2.12.

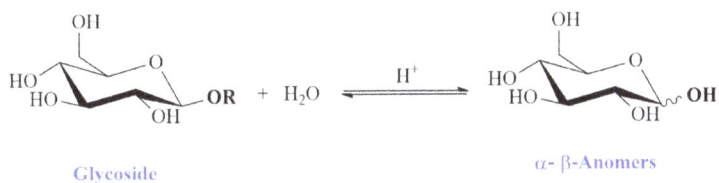

Glycoside

α- β-Anomers

Figure 2.12: Hydrolysis of glycosides.

Disaccharides can be readily hydrolyzed under weak acidic conditions, such as dilute HCl, producing their constitutive monomers in equivalent quantities (Figure 2.13).

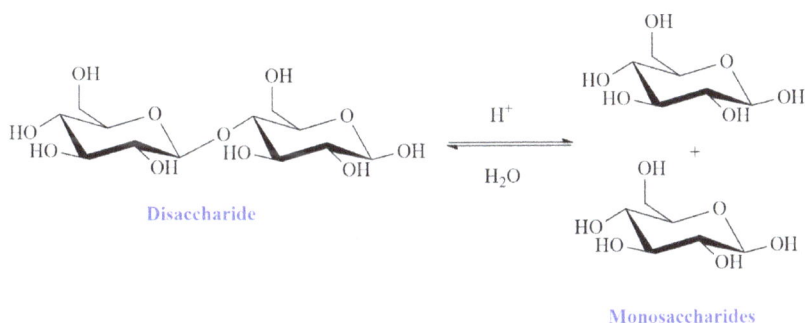

Disaccharide

Monosaccharides

Figure 2.13: Hydrolysis of disaccharides.

2.3.1.6.3 Alkylation of Glycosides (Formation of Ethers)

When glucoside reacts with methyl iodide in the presence of silver oxide, the methylated glycoside is obtained (Figure 2.14).

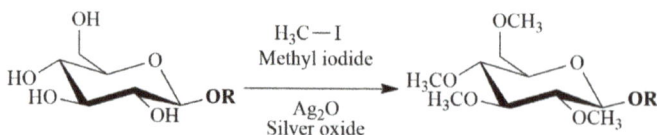

Figure 2.14: Formation of ethers (alkylation of glycosides).

2.3.1.6.4 Acylation of Glycosides (Formation of Esters)

When glucoside reacts with acetic anhydride in the presence of pyridine, the acetylated glycoside is obtained (Figure 2.15).

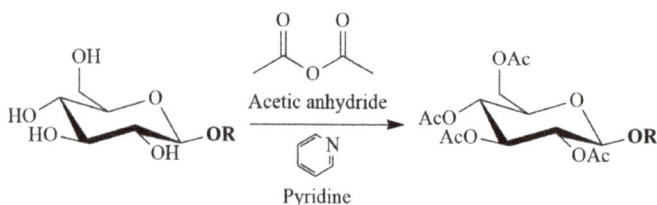

Figure 2.15: Formation of esters (acylation of glycosides).

2.3.1.6.5 Oxidation to Acidic Sugars

Oxidation of the aldehyde end of D-glucose with a weak oxidizing agent such as bromine water produces D-gluconic acid (Figure 2.16).

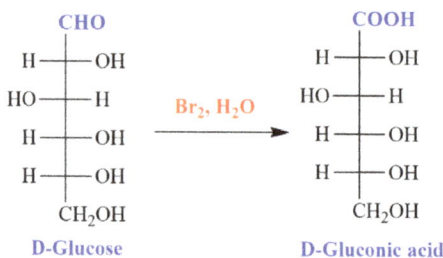

Figure 2.16: Oxidation of D-glucose to D-gluconic acid.

2.3.1.6.6 Reduction to Sugar Alcohols

The reduction of the carbonyl group with either an aldose or a ketose to a hydroxyl group, using hydrogen as the reducing agent or sodium borohydride, produces the corresponding polyhydroxy alcohol, which is sometimes called a *sugar alcohol*. For example, the reduction of D-glucose gives D-glucitol (Figure 2.17).

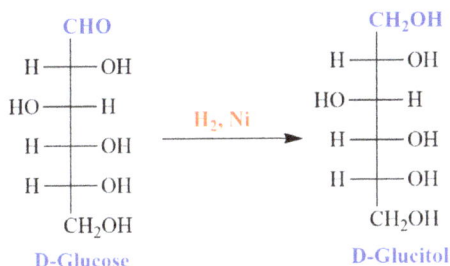

Figure 2.17: Reduction of D-glucose to D-glucitol.

2.3.1.7 Structural Variation in Monosaccharides

One or more -OH groups can be replaced by other atoms or groups leading to substituted sugars (Figure 2.18). Commonly occurring substituents are:

- H deoxy sugar – NH_2 amino sugar
- CH_2OH (hydroxyl methyl) – CH_3 (methyl)
- F fluoro sugar – SH thiosugar

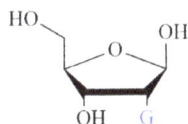

$G= H, F, NH_2, SH, CH_3, CH_2OH$ **Figure 2.18:** Substituted sugars.

2.3.1.8 Mutarotation

Mutarotation is the change in the equilibrium between two anomers due to optical rotation. Cyclic sugars show mutarotation as α and β anomeric forms interconvert. For example, an aqueous solution of D-glucose contains an equilibrium mixture of α-D-glucopyranose, β-D-glucopyranose, and the intermediate open-chain form (Figure 2.19).

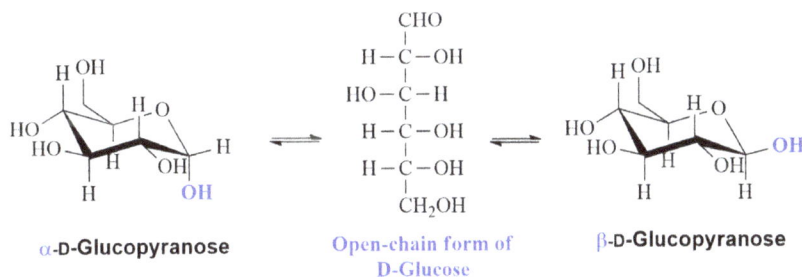

α-**D-Glucopyranose**   Open-chain form of   β-**D-Glucopyranose**
                                  D-Glucose

Figure 2.19: Mutarotation of D-glucose.

2.3.1.9 Epimerization

Epimerization is a process in stereochemistry of carbohydrates in which there is a change in the configuration of only one chiral center, for example, *C2-epimerization* between D-glucose and D-mannose and *C4-epimerization* between D-glucose and D-galactose (Figure 2.20).

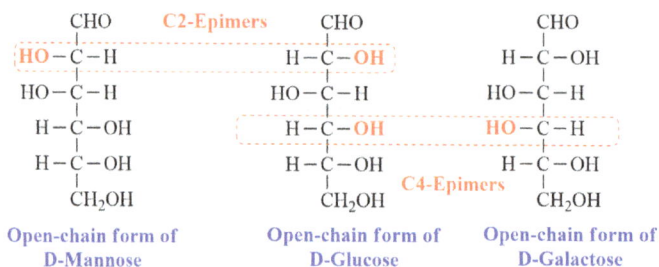

Figure 2.20: C2- and C4-epimerization between D-mannose, D-glucose, and D-galactose.

2.3.2 Disaccharides

2.3.2.1 Lactose

Lactose is a disaccharide that has galactose and glucose units connected through β (1→4) glycosidic linkage (Figure 2.21).

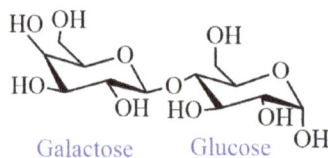

Figure 2.21: Chemical structure of lactose.

2.3.2.2 Sucrose

Sucrose is a disaccharide that has fructose and glucose units bonded through α, β (1→2) glycosidic linkage (Figure 2.22). Sucrose or saccharose is made by refining sugarcane or sugar beets.

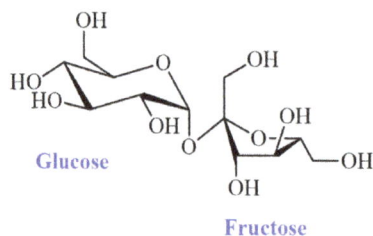

Figure 2.22: Chemical structure of sucrose.

2.3.2.3 Maltose

Maltose, or malt sugar, is a disaccharide that has two units of glucose joined by α (1→4) glycosidic bond (Figure 2.23).

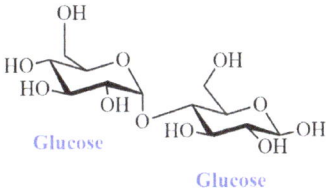

Figure 2.23: Chemical structure of maltose.

2.3.2.4 Cellobiose

Cellobiose is a disaccharide with two units of glucose joined by β (1→4) glycosidic bond and can be produced by the hydrolysis of cellulose (Figure 2.24).

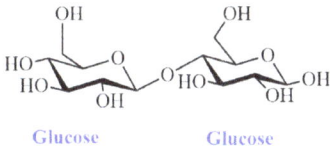

Figure 2.24: Chemical structure of cellobiose.

2.3.3 Oligosaccharides

2.3.3.1 Melitose

Melitose or raffinose is a trisaccharide that consists of a galactose unit connected to sucrose (glucose + fructose) by α (1→6) glycosidic bond (Figure 2.25).

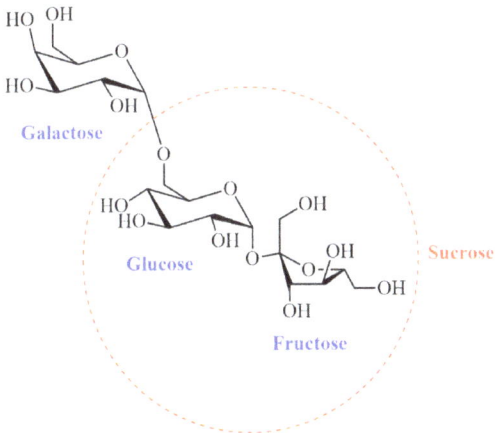

Figure 2.25: Chemical structure of melitose.

2.3.3.2 Maltotriose

Maltotriose is a trisaccharide that consists of three glucose units linked by α (1, 4) bonds (Figure 2.26).

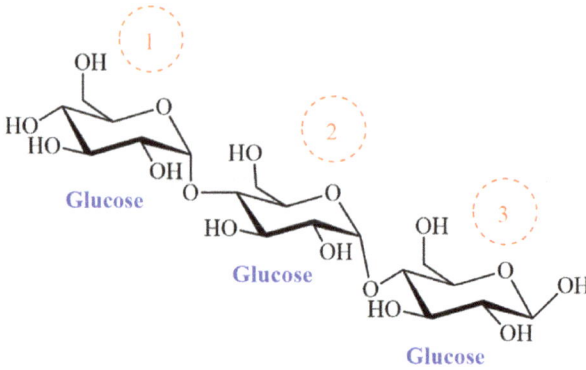

Figure 2.26: Chemical structure of maltotriose.

2.3.4 Polysaccharides

Polysaccharides are polymers that have many monosaccharide units bonded together by glycosidic linkages. They can simply be distinguished from each other by the type of the monosaccharide repeating unit(s) in the polymer chain or the length of the polymer chain or the type of glycosidic linkage between monomer units and the degree of branching of the polymer chain.

2.3.4.1 Starch

Starch has two types of polymers amylose and amylopectin, and both are made of α-D-glucose. *Amylose* is a linear polysaccharide (Figure 2.27).

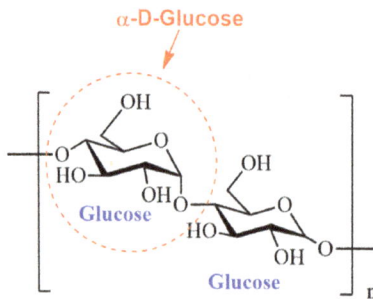

Figure 2.27: Chemical structure of amylose.

Amylopectin is a highly branched polysaccharide that differs from amylose in being highly branched. Short side chains of glucose units are attached with α (1→6) linkage approximately every 20–30 glucose units along the chain (Figure 2.28).

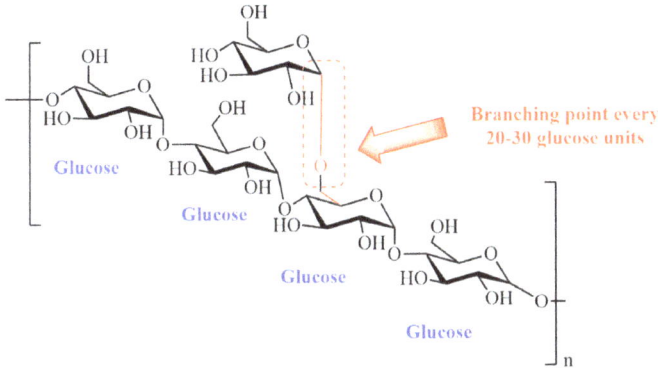

Figure 2.28: Chemical structure of amylopectin.

2.3.4.2 Cellulose

Cellulose is a polymer made of β-D-glucose units. Its hydroxymethyl (CH$_2$OH) groups are alternating above and below the plane of the cellulose molecule producing long unbranched chains (Figure 2.29). It exists in different forms such as microcrystalline cellulose, powdered cellulose, and low-crystallinity powdered cellulose.

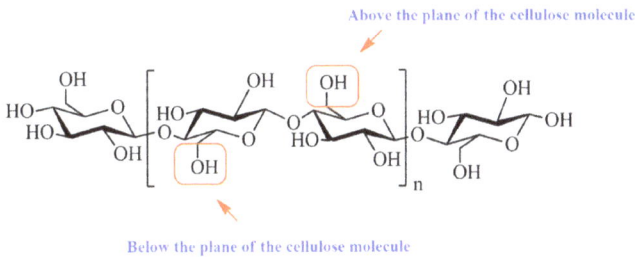

Figure 2.29: Cellulose chemical structure.

2.3.4.2.1 Cellulose Derivatives

Cellulose esters are water-insoluble polymers and are widely used in pharmaceutical controlled-release preparations. They are categorized into organic and inorganic groups. Examples of organic cellulose esters are cellulose acetate and cellulose acetate phthalate, and inorganic cellulose esters are cellulose nitrate (or pyroxylin) and cellulose sulfate (Figure 2.30).

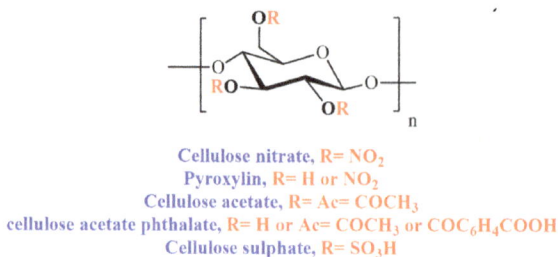

Cellulose nitrate, R= NO_2
Pyroxylin, R= H or NO_2
Cellulose acetate, R= Ac= $COCH_3$
cellulose acetate phthalate, R= H or Ac= $COCH_3$ or COC_6H_4COOH
Cellulose sulphate, R= SO_3H

Figure 2.30: Cellulose esters.

Cellulose ethers are high-molecular-weight compounds produced by replacing the hydrogen atoms of hydroxyl groups with alkyl or substituted alkyl groups. Examples of mostly used cellulose ethers are methylcellulose, ethyl cellulose, hydroxyethylcellulose, hydroxypropylcellulose, hydroxypropylmethylcellulose, carboxymethylcellulose, and sodium carboxymethyl cellulose (Figure 2.31).

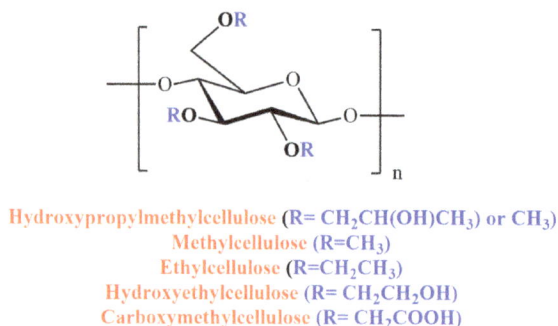

Hydroxypropylmethylcellulose (R= $CH_2CH(OH)CH_3$) or CH_3)
Methylcellulose (R=CH_3)
Ethylcellulose (R=CH_2CH_3)
Hydroxyethylcellulose (R= CH_2CH_2OH)
Carboxymethylcellulose (R= CH_2COOH)

Figure 2.31: Cellulose ethers.

2.3.4.2.2 Cellulose Applications
- Taste-masking agents
- Binders in the granulation process
- Thickening and stabilizing agents
- Pharmaceutical coating processes

2.3.4.3 Pullulan
Pullulan consists of maltotriose (three glucose molecules linked by α (1, 4) bonds) units linked through α (1, 6) glycosidic bonds (Figure 2.32). Pullulan is water-soluble, odorless, flavorless, edible, and makes strong films with high adhesion and oxygen barrier properties. Its films have low permeability to oxygen, which protects active ingredients, flavors, and colors incorporated into the film from

deterioration. Pullulan is used as a food ingredient and coating agent in food formulation and packaging industry.

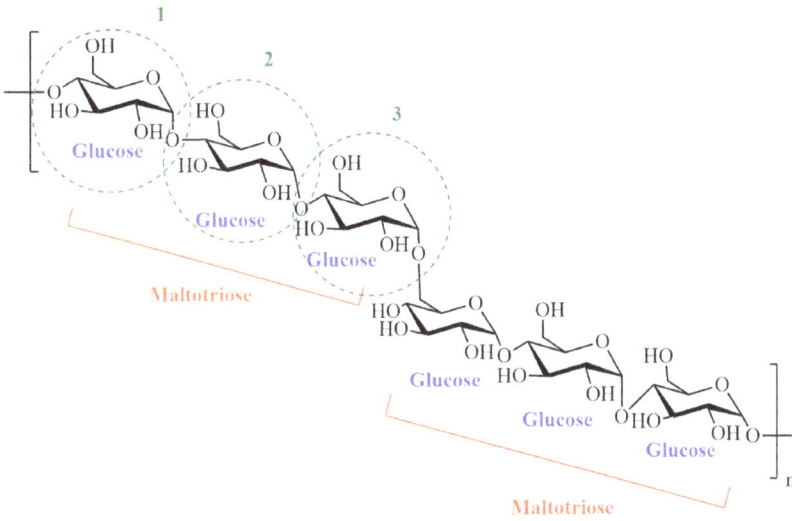

Figure 2.32: Chemical structure of pullulan.

2.3.4.4 Glycogen
Glycogen is a polymer of α-D-glucose. The branches in glycogen tend to be about 10–15 glucose units and shorter than those in the amylopectin (Figure 2.33).

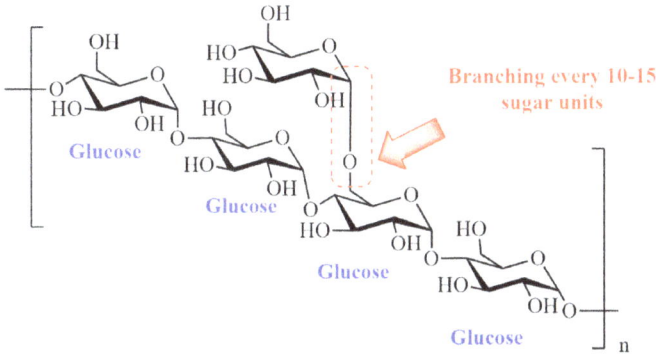

Figure 2.33: Chemical structure of glycogen.

2.3.4.5 Dextran
In dextran, the main chains are formed by α (1→6) glycosidic linkage, and the side branches are attached by α (1→2) or α (1→3) or α (1→4) linkages (Figure 2.34).

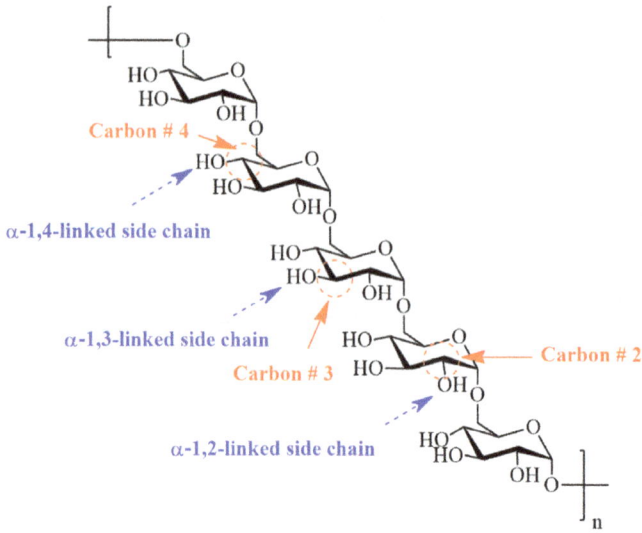

Figure 2.34: Chemical structure of dextran.

2.3.4.6 Inulin

Inulin is a water-soluble polysaccharide and its polymer consisting of fructose units linked via β (2→1) linkages that have a terminal glucose unit (Figure 2.35).

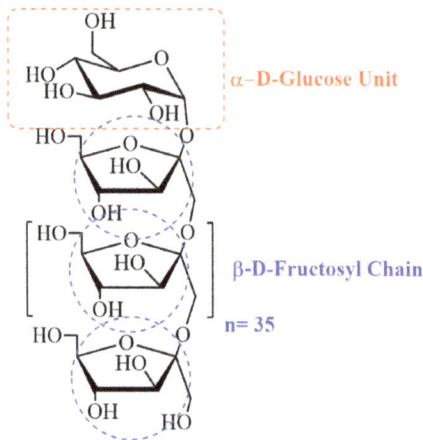

Figure 2.35: Chemical structure of inulin.

2.3.4.7 Chitin

Chitin is a polysaccharide that is similar to cellulose in both function and structure. Structurally, chitin is a linear polymer (no branching) with all β (1→4) glycosidic linkages, as is cellulose. Chitin differs from cellulose in that the monosaccharide present is an *N*-acetyl amino derivative of D-glucose (Figure 2.36).

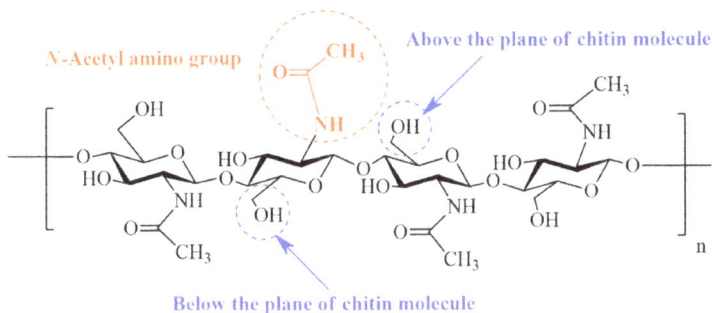

Figure 2.36: Chitin chemical structure.

2.4 Summary and Important Concepts

Carbohydrates. Carbohydrates are polyhydroxyaldehydes and/or polyhydroxyketones.

Carbohydrate classification. Carbohydrates are classified into four groups: monosaccharides, disaccharides, oligosaccharides, and polysaccharides.

Chirality and achirality. A chiral object is not identical to its mirror image. An achiral object is identical to its mirror image.

Chiral or stereogenic center. Chiral or stereogenic center is a carbon atom that has four different groups bonded to it.

Chirality of monosaccharides. Monosaccharides are classified as D or L sugars on the basis of the configuration of the chiral center farthest from the carbonyl group, or in other words, the configuration at the highest stereogenic center.

Classification of monosaccharides. Monosaccharides are classified as aldoses or ketoses on the basis of the type of carbonyl group present. They are further classified as trioses, tetroses, pentoses, and hexoses on the basis of the number of carbon atoms present in the open-chain structure.

Alditol. Alditol is an alcohol that results from the reduction of the aldehyde or keto group of an aldose or ketose. An example is reduction of glucose to glucitol.

Important monosaccharides. Important monosaccharides include glucose, galactose, fructose, and ribose. Glucose and galactose are aldohexoses, fructose is a ketohexose, and ribose is an aldopentose.

Aldose. Aldose is a monosaccharide containing an aldehyde group.

Ketose. Ketose is a monosaccharide containing ketone group.

Anomers. Anomers are diastereomers that differ only in the configuration at the acetal or hemiacetal carbon of a sugar cyclic form.

Furanose. Furanose is a sugar in which the cyclic acetal or hemiacetal ring is five membered.

Cellulose applications. Cellulose is used as a taste-masking agent, a binder in the granulation process, a thickening and stabilizing agent, and in the pharmaceutical coating processes.

Reactions of monosaccharides. Among important reactions of monosaccharides are the oxidation to an acidic sugar, reduction to a sugar alcohol, and glycoside formation.

Disaccharides. Disaccharides are glycosides formed from the linkage of two monosaccharides.

Polysaccharide. Polysaccharide comprises monosaccharides joined together by glycosidic linkages.

Mutarotation. Mutarotation is the change in the equilibrium between two anomers due to the optical rotation.

Epimerization. Epimerization is a process in stereochemistry of carbohydrates in which there is a change in the configuration of only one chiral center.

2.5 Practice Exercises

2.5.1 Why is D-glucose classified and named as aldohexose sugar?

2.5.2 List down the name of the main sugar units in raffinose.

2.5.3 What is the main difference between amylopectin and glycogen?

2.5.4 Why is sucrose considered a disaccharide? What are the sugar units in lactose?

2.5.5 Classify carbohydrates on the basis of sugar units' numbers.

2.5.6 Draw the chemical structure of any given ketohexose sugar and number the carbon atoms on the structure.

2.5.7 What are the common names of D-fructose?

2.5.8 Show how D-glucopyranose is formed from D-glucose.

2.5.9 Draw the chemical structures of ribose and deoxyribose sugars.

2.5.10 What type of glycosidic linkages do sucrose, lactose, and maltose have?

2.5.11 List down two commonly occurring substituents in the structural variation in carbohydrates.

2.5.12 Draw a Fischer projection formula for the enantiomer of the following monosaccharide:

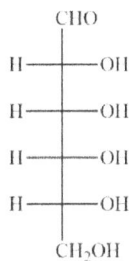

```
        CHO
        |
H———————OH
        |
H———————OH
        |
H———————OH
        |
H———————OH
        |
      CH₂OH
```

2.5.13 Classify each of the following monosaccharides as a D enantiomer or an L enantiomer:

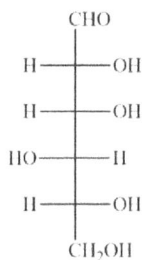

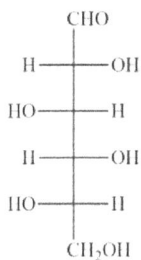

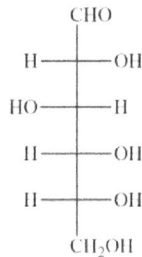

```
       CHO                 CHO                 CHO
        |                   |                   |
H——————OH            H——————OH            H——————OH
        |                   |                   |
H——————OH            HO—————H            HO—————H
        |                   |                   |
HO—————H             H——————OH            H——————OH
        |                   |                   |
H——————OH            HO—————H            H——————OH
        |                   |                   |
     CH₂OH               CH₂OH               CH₂OH
```

2.5.14 Classify each of the following monosaccharides according to both the number of carbon atoms and the type of carbonyl group present:

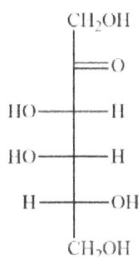

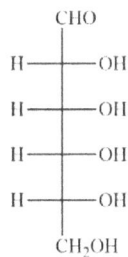

```
      CH₂OH                  CHO
        |                     |
        =O            H——————OH
        |                     |
HO——————H             H——————OH
        |                     |
HO——————H             H——————OH
        |                     |
H———————OH            H——————OH
        |                     |
      CH₂OH                CH₂OH
```

2.5.15 Indicate how many sugar units are present in each of the following:
i.   Disaccharide
ii. Oligosaccharide
iii. Polysaccharide

2.5.16 How many carbon atoms and oxygen atoms exist in the ring portion of the cyclic forms of each of the following monosaccharides?

i.   D-Glucose

ii. D-Galactose

iii. D-Fructose

2.5.17 Which of the following monosaccharides is a *reducing sugar* and why?

i.   D-Glucose

ii. D-Galactose

iii. D-Fructose

2.5.18 What kind of glycosidic linkage is present in each of the following disaccharides?

i.   Sucrose

ii. Lactose

iii. Cellobiose

2.5.19 Indicate whether or not each of the following is a correct characterization for the amylopectin form of starch.

i.   It is a polysaccharide.

ii. It contains two different types of monosaccharides.

iii. It is a branched-chain glucose polymer.

2.5.20 What monosaccharide(s) is (are) obtained from the hydrolysis of each of the following?

i.   Sucrose

ii. Glycogen

iii. Starch

2.5.21 Classify each of the following polysaccharides as a glucose polymer or a glucose-derivative polymer:

i.   Chitin

ii. Amylopectin

iii. Glycogen

2.5.22 Name the sugar units in the following cellobiose structure:

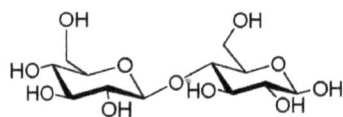

2.5.23 What is the main difference in structure between chitin and cellulose?

2.5.24 What does ketopentose stand for?

2.5.25 What types of glycosidic bonds do the following sugars have?
i. Maltose
ii. Sucrose
iii. Lactose

Chapter 3
Lipids

3.1 Types (classes) of Lipids

Lipids can be classified into waxes, triglycerides (triacylglycerols), phospholipids (glycerophospholipids), sphingolipids (glycolipids and sphingomyelins), eicosanoids (prostaglandins and leukotrienes), and steroids.

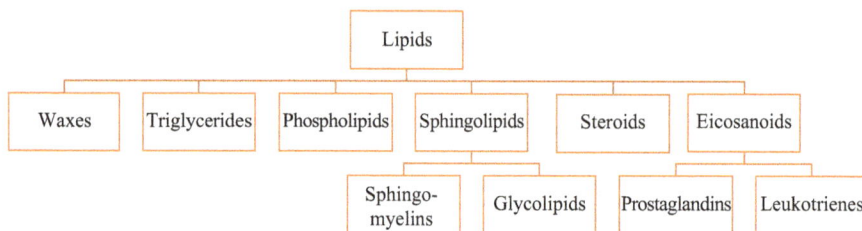

3.2 Fatty Acids

Fatty acids are long-chain carboxylic acids. Their size ranges from C_{12} to C_{26} carbon atoms for long-chain fatty acids, C_8 and C_{10} for medium-chain fatty acids, and C_4 and C_6 for short-chain fatty acids. Fatty acids do not dissolve in water but dissolve in fat solvents such as hexane, ether, acetone, and chloroform. They can be categorized as follows:

3.2.1 Saturated fatty acids are the ones that contain only C–C single bonds. They are closely packed and have strong attractions between chains. They also have high melting points and exist as solids at room temperature.

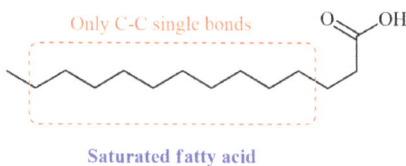

Saturated fatty acid

3.2.2 Unsaturated fatty acids are the second type of fatty acids. They have one or more C=C bonds and C–C bonds. Since they have few interactions between chains, they are not closely packed. They also have low melting points and exist as liquids at room temperature.

https://doi.org/10.1515/9783110762761-003

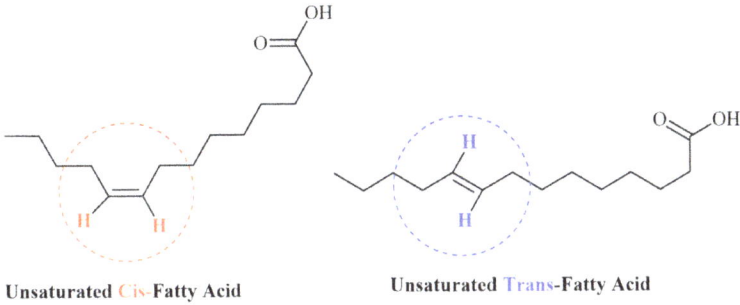

Unsaturated Cis-Fatty Acid

Unsaturated Trans-Fatty Acid

3.2.3 Some of the common and well-known fatty acids are listed in Table 3.1.

Table 3.1: Well-known fatty acids.

No. of carbon atoms/no. of double bonds	Common name
Saturated fatty acids	
12:0	Lauric acid
14:0	Myristic acid
16:0	Palmitic acid
18:0	Stearic acid
20:0	Arachidic acid
Monounsaturated fatty acids	
16:1 (Δ^9) ω-7	Palmitoleic acid
18:1 (Δ^9) ω-9	Oleic acid
Polyunsaturated fatty acids	
18:2 ($\Delta^{9,\ 12}$) ω-6	Linoleic acid
18:3 ($\Delta^{9,\ 12,\ 15}$) ω-3	Linolenic acid
20:4 ($\Delta^{5,\ 8,\ 11,\ 14}$) ω-6	Arachidonic acid

3.3 Waxes

Waxes are long-chain esters made from mixtures of long-chain carboxylic acids (**even no. 16–36**) and long-chain alcohols (**even no. 24–36**). The chemical structure for two well-known waxes (bees wax from bee honeycombs and carnauba wax from Brazilian palm tree leaves) is given in Figure 3.1.

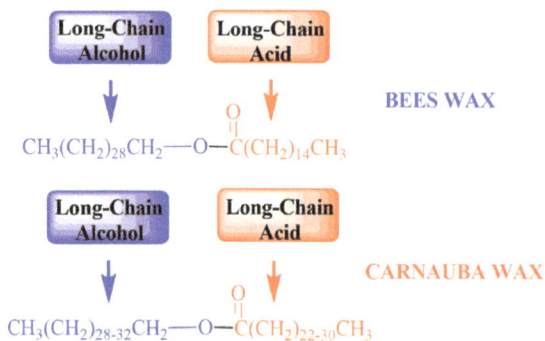

Figure 3.1: The chemical structures of beeswax and carnauba wax.

3.4 Fats and Oils (Triglycerides or Triacylglycerols)

All fats and oils are composed of triesters of glycerol (1, 2, 3-propanetriol, also known as glycerin with three fatty acids). They are named chemically as triacylglycerols but are often called triglycerides (Figure 3.2). The three fatty acids of any specific triacylglycerol are not necessarily the same.

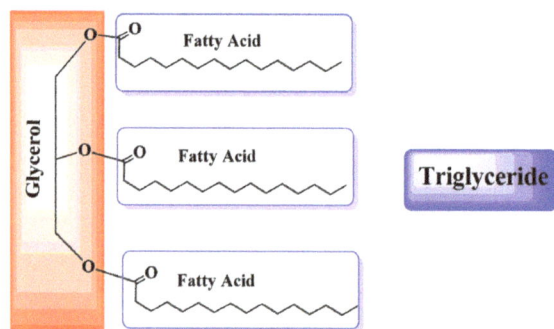

Figure 3.2: The chemical structure of triglyceride.

Triglycerides are formed by the esterification of glycerol with three fatty acids as shown in Figure 3.3.

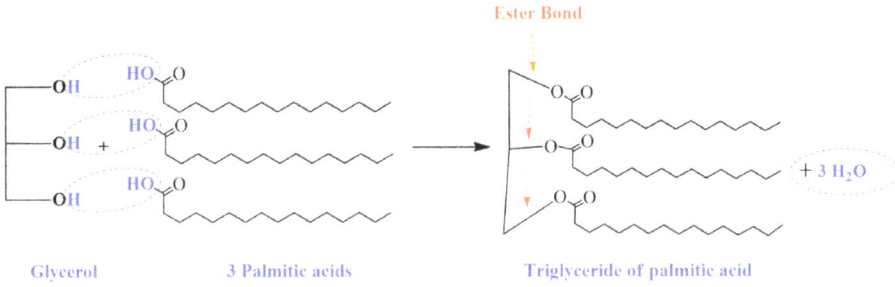

Figure 3.3: Triglyceride formation.

3.4.1 Chemical Reactions of Triglycerides

3.4.1.1 Hydrogenation: unsaturated triglycerides react with hydrogen (H_2) in the presence of nickel (Ni) or platinum (Pt) catalysts to form saturated triglycerides. The C=C bonds are converted to C–C bonds (Figure 3.4).

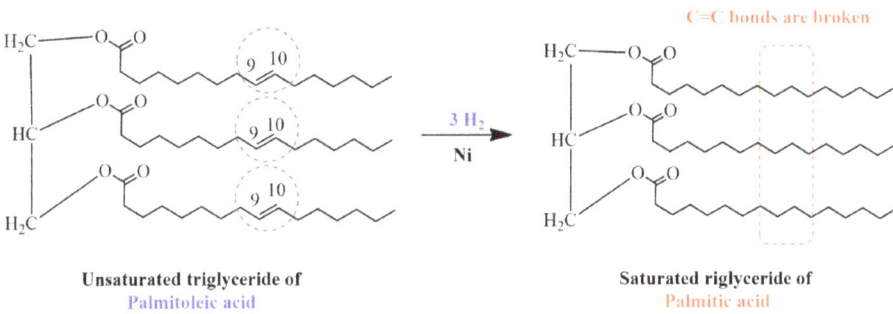

Figure 3.4: Hydrogenation of unsaturated triglycerides.

3.4.1.2 Hydrolysis: triglycerides split by water and acid or enzyme catalyst to glycerol and three fatty acids. They also split into glycerol and the salts of fatty acid "soaps" by the action of strong base such as KOH or NaOH in a reaction called saponification as shown in Figure 3.5.

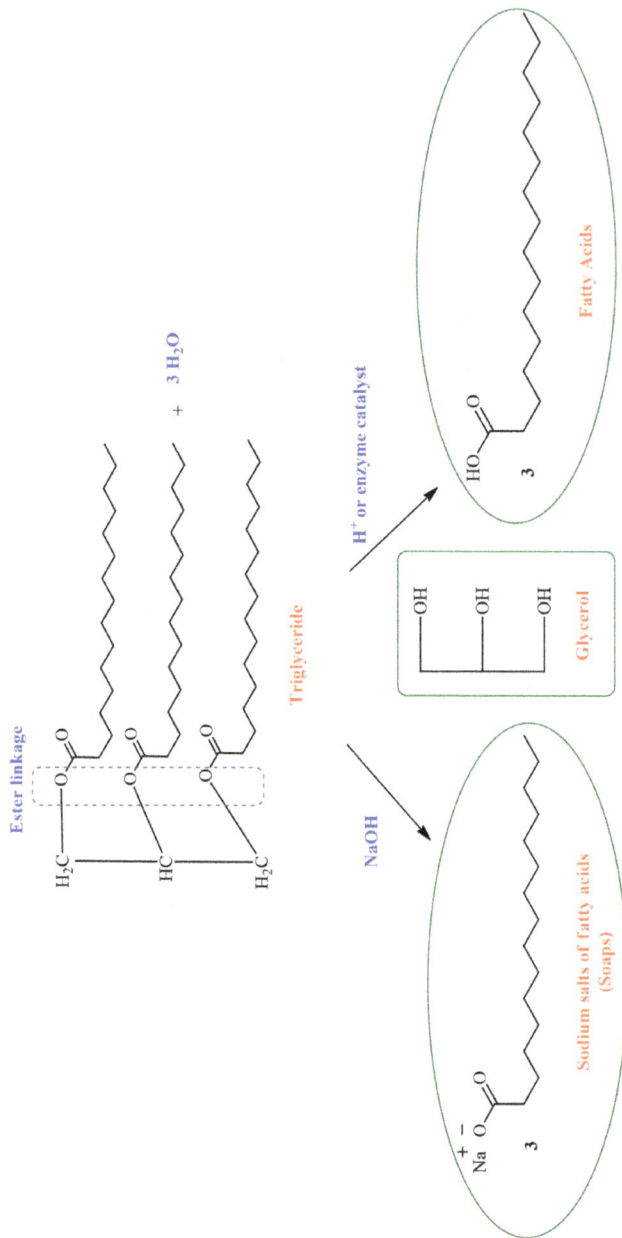

Figure 3.5: Hydrolysis of triglycerides.

3.5 Phospholipids

Phospholipids, also known as glycerophospholipids, are derived from the phosphatidic acid (Figure 3.6). It is a molecule in which glycerol is esterified with two molecules of fatty acids (*two out of the following: palmitic acid, stearic acid, and oleic acid*) and one of phosphoric acid.

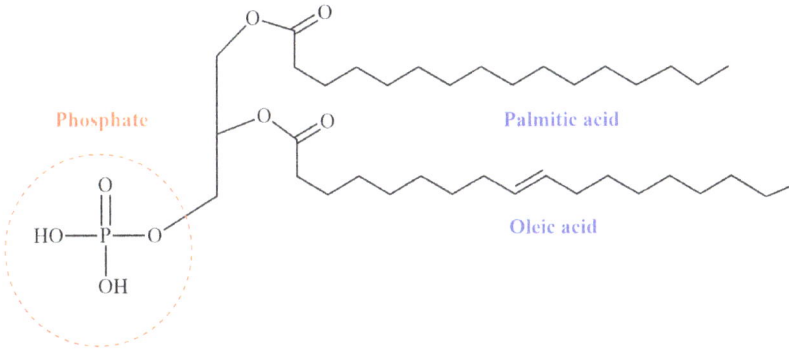

Figure 3.6: Phosphatidic acid.

Further esterification of phosphatidic acid with low-molecular-weight amino alcohols/alcohols gives phospholipids as shown in Figure 3.7.

3.6 Sphingolipids

There are two types of sphingolipids: the first one is a sphingophospholipid which contains one fatty acid and one phosphate group with an amino alcohol attached to a sphingosine. The other one is a sphingoglycolipid, which contains both a fatty acid and a carbohydrate (galactose or glucose) component attached to a sphingosine molecule (Figure 3.8).

3.7 Eicosanoids: Prostaglandins and Leukotrienes

Eicosanoids are a group of 20-carbon carboxylic acids. These lipids are synthesized in nature from arachidonic acid (eicosanoic acids). Prostaglandins and leukotrienes are synthesized in the body from the 20-carbon unsaturated fatty acids, namely, arachidonic acid. They are one of two classes of eicosanoids. All prostaglandins contain a five-membered ring, which the leukotrienes lack (Figure 3.9).

Palmitic acid

Oleic acid

Phosphate

G =

Alcohol Name	Choline	Ethanolamine	Serine	Inositol (myo-inositol)
Phospholipid Name	Phosphatidylcholine (Lecithin)	Phosphatidylethanolamine (Cephalin)	Phosphatidylserine	Phosphatidylinositol

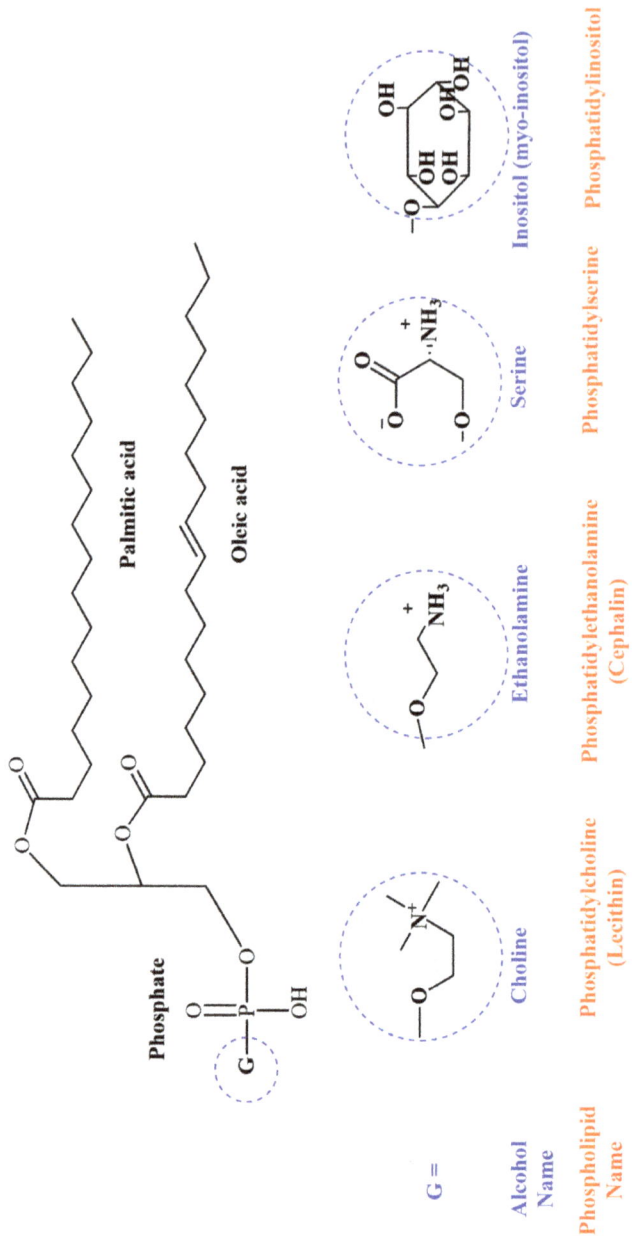

Figure 3.7: Examples of phospholipids.

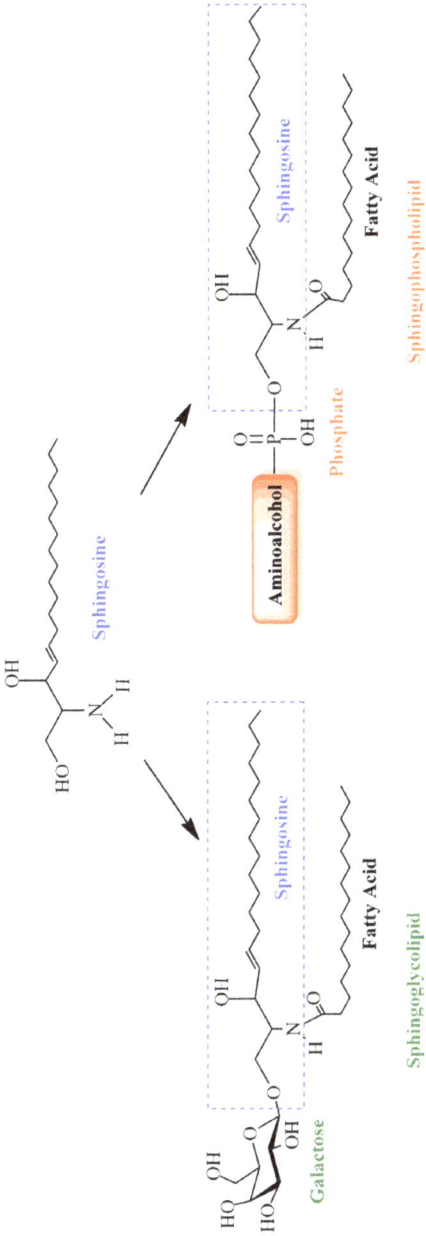

Figure 3.8: Types of sphingolipids.

Arachidonic Acid (Bent)

Leukotriene

Prostaglandin

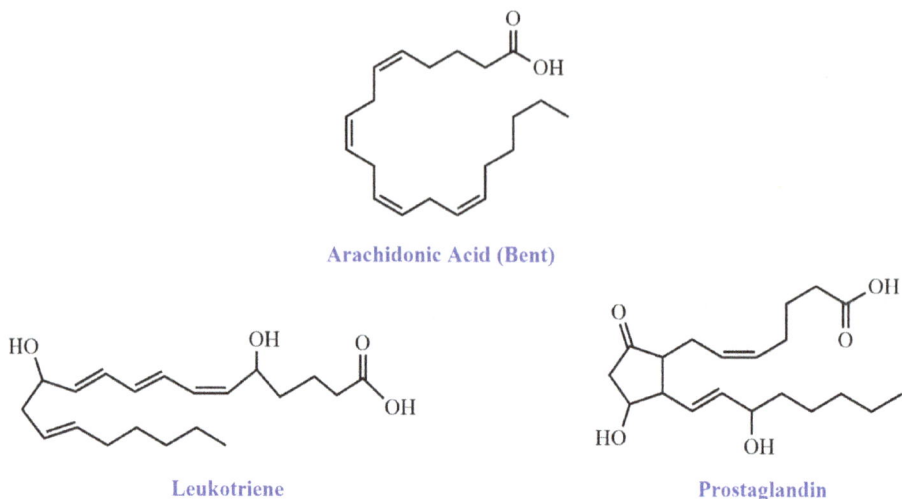

Figure 3.9: Chemical structure of eicosanoids and arachidonic acid.

3.8 Steroids

The structures of steroids are based on a fused tetracyclic ring system that involves three 6-membered rings and one 5-membered ring. The four rings are designated A, B, C, and D, beginning at the lower left. A good example for steroids is cholesterol (Figure 3.10). It has a molecular formula of $C_{27}H_{45}OH$ and composed of three regions: a hydrocarbon tail, a ring structure region with four hydrocarbon rings, and a hydroxyl group.

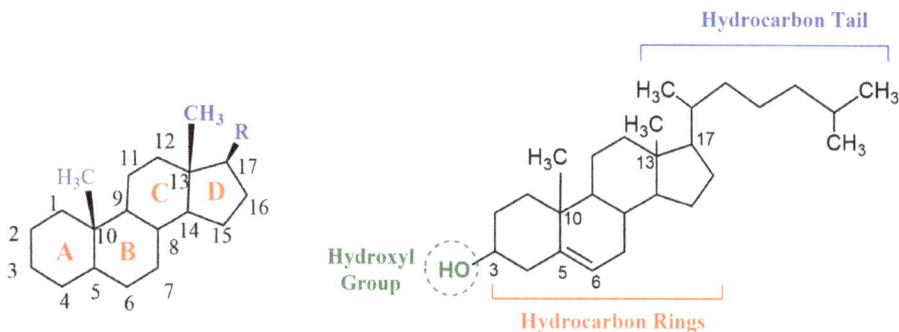

Figure 3.10: Steroid structure and cholesterol.

3.9 Terpenes

Terpenes are compounds of five-carbon isopentyl (isoprene) groups. They are composed of two or more isoprene units. Terpenes (Figure 3.11) are classified by the number of carbons they contain in groups of 10:

- A monoterpene has 10 carbons and 2 isoprene units.
- A sesquiterpene has 15 carbons and 3 isoprene units.
- A diterpene has 20 carbons and 4 isoprene units.

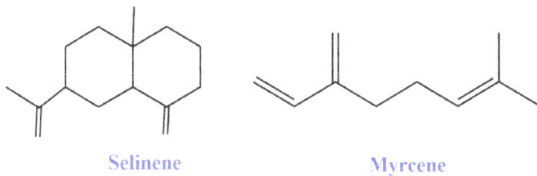

Selinene   Myrcene

Figure 3.11: Examples of terpenes.

3.10 Vitamins

A vitamin is an organic compound that has important functions in the human body and must be obtained from dietary sources in our meals. Based on solubility, vitamins can be classified into two major classes: the water-soluble vitamins and the fat (lipid)-soluble vitamins. There are nine water-soluble vitamins (Figure 3.12) and four fat-soluble vitamins (Figure 3.13).

3.10.1 Water-Soluble Vitamins

- Thiamin (vitamin B1)
- Riboflavin (vitamin B2)
- Niacin (vitamin B3)
- Vitamin B6
- Folate
- Vitamin B12
- Pantothenic acid (vitamin B5)
- Biotin
- Vitamin C

3.10.1.1 General properties of water-soluble vitamins
- They do not require carriers.
- They are absorbed directly into the blood.

- Their excess can be safely removed by the body through the kidneys.
- Circulate in the water-containing parts of the human body.

3.10.1.2 Chemical Structures of Selected Water-Soluble Vitamins

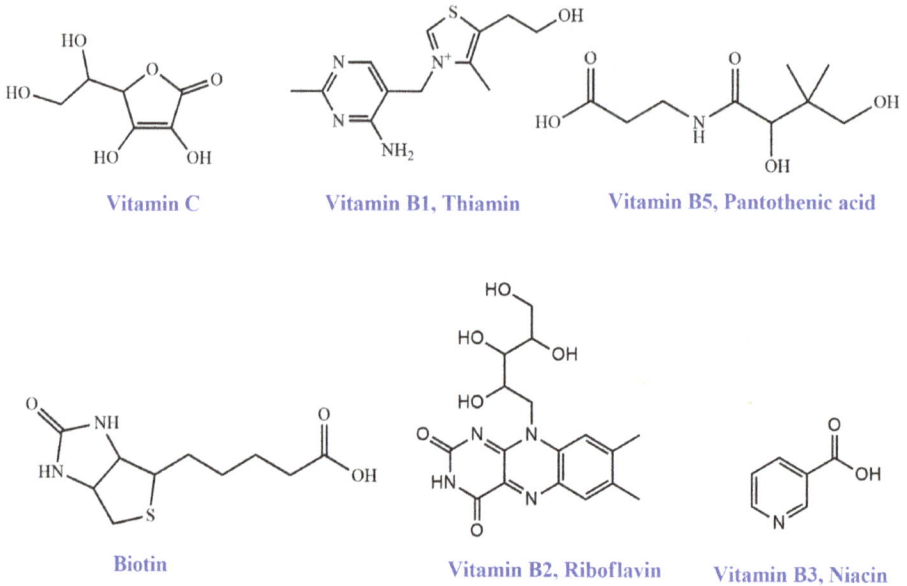

Vitamin C   Vitamin B1, Thiamin   Vitamin B5, Pantothenic acid

Biotin   Vitamin B2, Riboflavin   Vitamin B3, Niacin

Figure 3.12: Chemical structures of selected water-soluble vitamins.

3.10.2 Fat-Soluble Vitamins

- Vitamin A
- Vitamin D
- Vitamin E
- Vitamin K

3.10.2.1 General Properties of Fat-Soluble Vitamins
- Many of them require protein carriers.
- They enter to the lymph system first.
- They remain in fat storage sites.
- They are found in fat cells.

3.10.2.2 Chemical Structures of Selected Fat-Soluble Vitamins

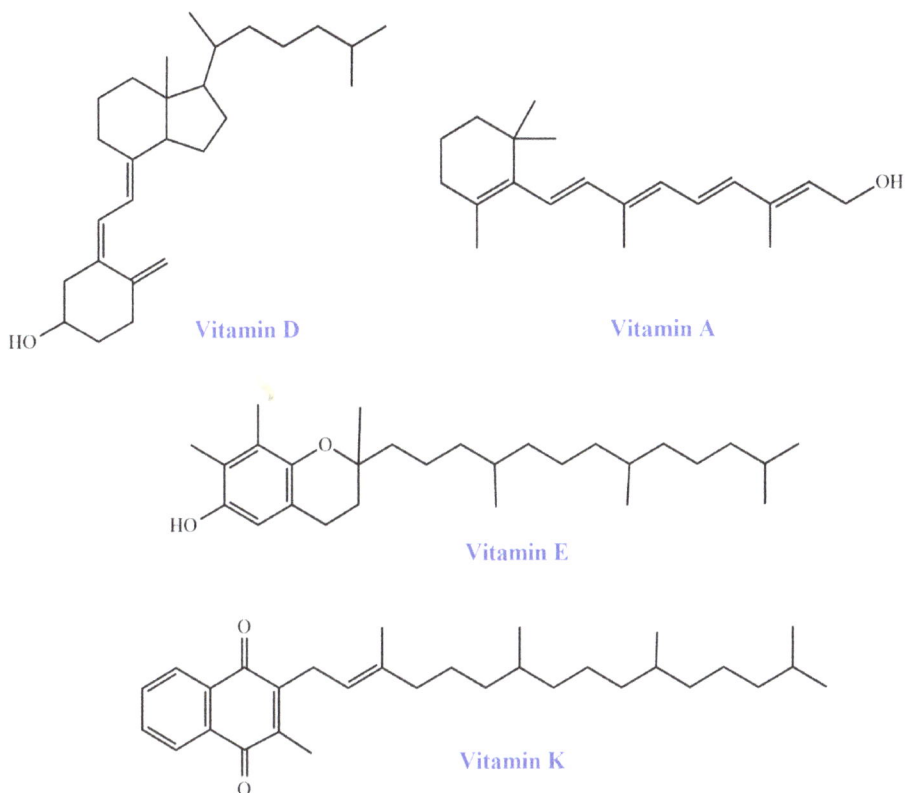

Figure 3.13: Chemical structures of selected fat-soluble vitamins.

3.11 Summary and Important Concepts

Lipids. Lipids are nonpolar solvent substances, which include fatty acids, triglycerides, waxes, prostaglandins, steroids, and terpenes.

Fatty acids. Fatty acids are long-chain carboxylic acids. They contain even numbers of carbon atoms between 12 and 26. Their sizes range from C_{12} to C_{26} carbon atoms for long-chain fatty acids, C8 and C10 for medium-chain fatty acids, and C_4 and C_6 for short-chain fatty acids.

Saturated fatty acids. Saturated fatty acids contain only carbon–carbon single bonds.

Unsaturated fatty acids. Unsaturated fatty acids have one or more C=C bonds in addition to C–C bonds.

Glycerides. Glycerides are fatty acid esters of glycerol.

Lipids. Lipids are substances that are not soluble in water and can be extracted from tissues by nonpolar organic solvents called fat solvents.

Triglyceride. Triglyceride is a fatty acid triester of glycerol.

Phosphoglyceride. Phosphoglyceride is an ester of glycerol in which the three hydroxy groups are esterified by two fatty acids and one phosphoric acid derivative.

Phosphatidic acid. Phosphatidic acid is a molecule in which glycerol is esterified with two molecules of fatty acids and one phosphoric acid.

Cephalins (phosphatidyl ethanolamines). Cephalins are phosphoglycerides with ethanolamine esterified to the phosphoric acid group.

Lecithins (phosphatidylcholines). Lecithins are phosphoglycerides with choline esterified to the phosphoric acid group.

Phospholipids (glycerophospholipids). Phospholipids are lipids derived from phosphatidic acid.

Wax. Wax is an ester of long-chain fatty acid with long-chain alcohol.

Steroids. Steroids are fused tetracyclic ring systems that involve three 6-membered rings and one 5-membered ring.

Eicosanoids. These are a group of 20-carbon carboxylic acids.

Saponification. This is the base-promoted hydrolysis of an ester, which is originally used to describe the hydrolysis of fats to make soap.

Soap. This is the sodium or potassium salts of fatty acids.

Terpenes. Terpenes are a family of compounds with carbon skeletons composed of two or more 5-carbon isoprene units.

Terpenoids. Terpenoids are a family of compounds including both terpenes and compounds of terpene origin whose carbon skeletons have been altered.

Vitamins. A vitamin is an organic compound that has important functions in the human body and must be obtained from dietary sources in our meals. Based on solubility, vitamins can be classified into two major classes: the water-soluble vitamins and the fat (lipid)-soluble vitamins. There are nine water-soluble vitamins and four fat-soluble vitamins.

Water-soluble vitamins. Vitamin C and the eight B vitamins are water-soluble vitamins.

Fat-soluble vitamins. The four fat-soluble vitamins are vitamins A, D, E, and K.

3.12 Practice Exercises

3.12.1 Match the following statements in column **A** with the correct answer in column **B** provided in the following table:

Column A	Column B
1. The notation for palmitic acid is _____	A. A16:0
2. Fatty acids are _____	B. 14:0
3. The structures of steroids are based on	C. Long-chain carboxylic acids
_____	D. Hydrogenation catalysts
4. Eicosanoids are a group of _____	E. Twenty-carbon carboxylic acids
5. The long-chain alcohol portion of the beeswax	F. Fused tetracyclic ring system
chemical structure has _____	G. 30 carbons
6. Nickel (Ni) and platinum (Pt) are _____	H. 28 carbons
7. The notation for myristic acid is _____	I. Long-chain alcohols

3.12.2 Which one of the following fatty acid structures has the **cis** configuration? Why?

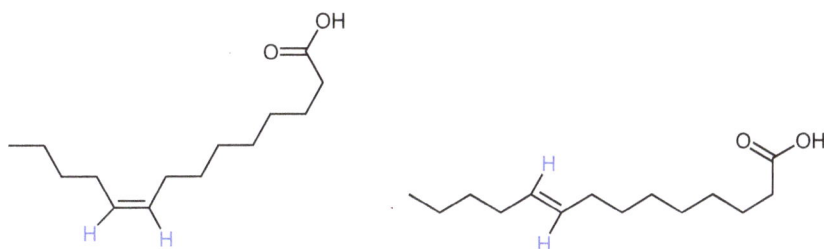

3.12.3 What does the following notation 16:1 (Δ^9) ω-7 mean?

3.12.4 Complete the missing information (**?**) in the following hydrogenation reaction.

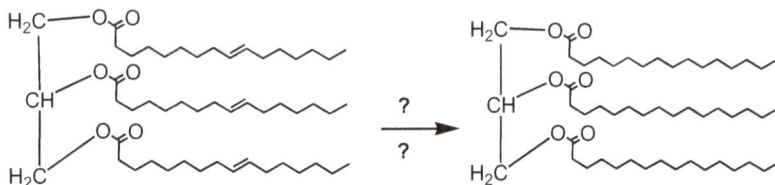

3.12.5 Assign numbers to the atoms of the following steroid structure.

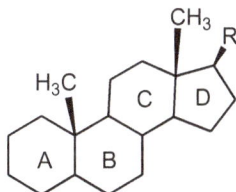

3.12.6 Draw the chemical structure of oleic acid and show the location of the double bond.

3.12.7 Write down the shorthand notations of the following fatty acids' structures:

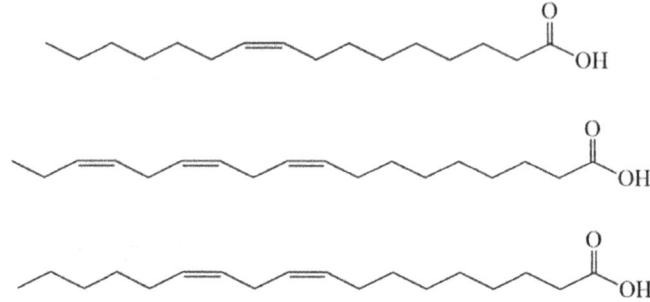

3.12.8 Show and name the three regions in the following cholesterol structure:

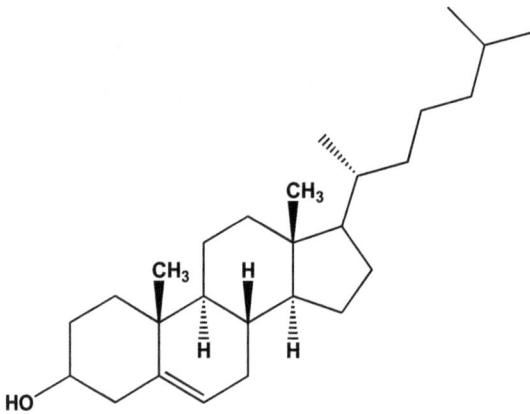

3.12.9 Draw the chemical structure of carnauba wax.

3.12.10 What are the three regions in the cholesterol?

3.12.11 What are the acid hydrolysis products of the triglycerides?

3.12.12 Is cholesterol a simple lipid or a complex lipid?

3.12.13 What group of lipids are esters of long-chain fatty acids and long-chain alcohols?

3.12.14 Is the structure shown below for wax or for triglyceride?

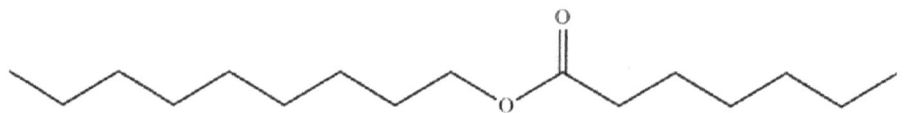

3.12.15 How many fatty acids are required to produce one molecule of a fat or oil?

3.12.16 Linoleic acid is an example of _____ fatty acid. It contains two carbon–carbon double bonds and _____ carbon atoms.

3.12.17 What are the two functional groups that chemically react to form triglyceride?

3.12.18 Ω-6 (ω-6) and Ω-3 (ω-3) fatty acids are important in the area of nutrition. Which of the following structures would be classified as an Ω-3 (ω-3) fatty acid?

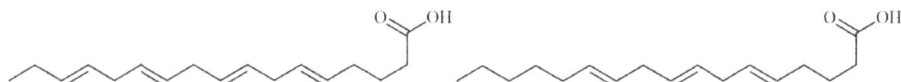

3.12.19 Which of the following fatty acids is omega-3 fatty acids?
i.   Oleic acid
ii. Linolenic acid
iii. Arachidonic acid

3.12.20 Draw the two fatty acid structures that result from acid and base hydrolysis of a triglyceride.

3.12.21 Which of the following terms best describes the structure below?

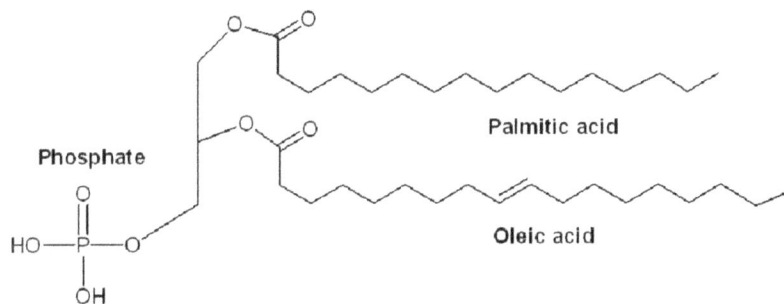

3.12.22 Label the hydrophilic and hydrophobic regions of the molecule shown below.

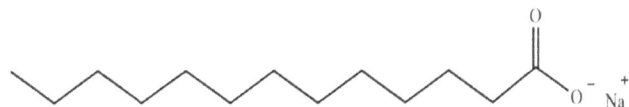

3.12.23 Of the two vitamins A and C, one is hydrophilic and water-soluble while the other is hydrophobic and fat-soluble. Which is which?

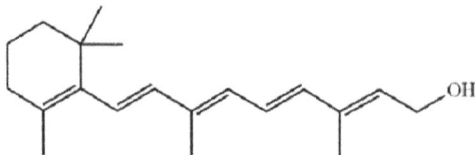 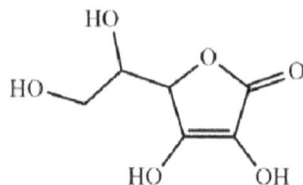

Vitamin A   **Vitamin C**

3.12.24 Which of the following is a second name for a phosphatidyl choline?
i.   Steroid
ii. Lecithin
iii. Cephalin

3.12.25 Draw the basic four-ring skeletal structure of steroids.

Chapter 4
Proteins (Polymers of Amino Acids)

4.1 Definition

Proteins and polypeptides are the polymers of amino acids. They can be made from several polypeptide chains as well as single chains. Their structure can also include nonpeptide components, such as saccharide chains and lipids. The amino acid residues in the proteins are joined together by peptide bonds.

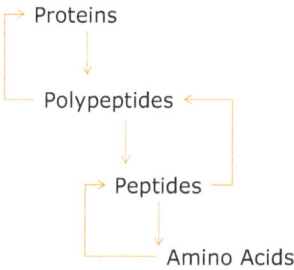

4.2 Amino Acids (Building Blocks of Proteins)

There are four different groups in the amino acid structures (except for glycine that has three different groups). Among these groups are the amino group ($-NH_2$), and the carboxyl group ($-COOH$) (Figure 4.1). The amino acids that are found in proteins have the amino group attached to the carbon atom next to the carboxyl group. They are known as α- or 2-amino acids.

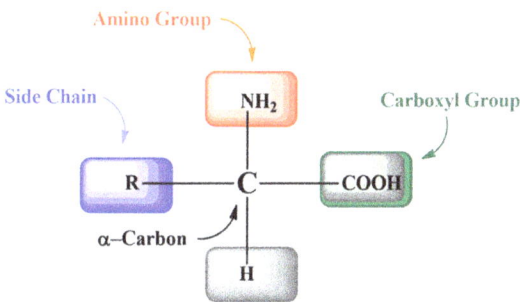

Figure 4.1: General structure of amino acids.

https://doi.org/10.1515/9783110762761-004

4.2.1 Structures and Nomenclature of Amino Acids

Amino acids can be distinguished from each other by the type of side-chain "R" present in their structure. It is not necessarily a simple hydrocarbon group like in alanine or single hydrogen atom in case of glycine. The side chain can contain other active groups like hydroxyl, thiol, amino, or carboxyl.

The simplest possible amino acid has a direct linkage of -NH$_2$ to -COOH, that is, H$_2$N-COOH, known as carbamic acid, and is highly unstable, but derivatives of carbamic acid, such as esters (H$_2$NCOOR), are sufficiently stable to be isolated. Amino acids can be grouped based on their side-chain polarity into four categories:

– **Nonpolar amino acids**

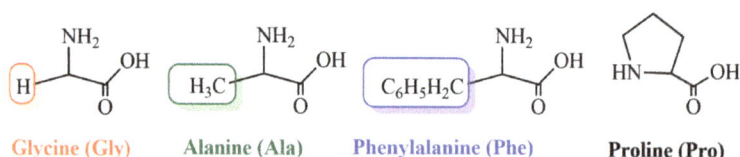

Glycine (Gly)   Alanine (Ala)   Phenylalanine (Phe)   Proline (Pro)

– **Polar neutral amino acids**

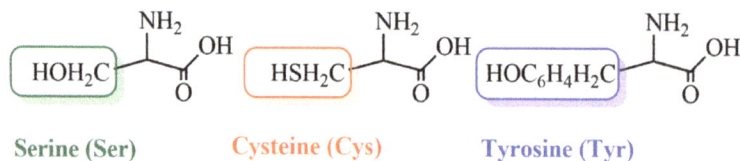

Serine (Ser)   Cysteine (Cys)   Tyrosine (Tyr)

– **Polar acidic amino acids**

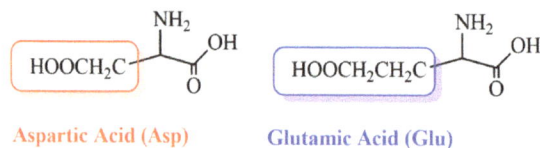

Aspartic Acid (Asp)   Glutamic Acid (Glu)

– **Polar basic amino acids**

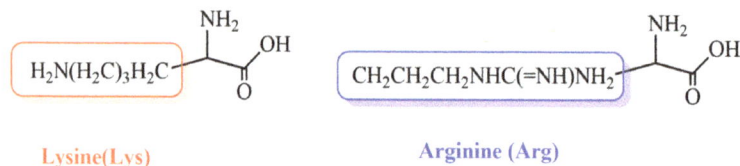

Lysine(Lys)   Arginine (Arg)

The 20 amino acids comprise the building blocks for the synthesis of proteins. The remaining two additional amino acids are derived by modification after biosynthesis of the protein. Hydroxyproline and cystine are synthesized from proline and cysteine, respectively, after the protein chain has been synthesized. Hydroxyproline is produced by hydroxylation of proline by the enzyme prolyl hydroxylase (Figure 4.2).

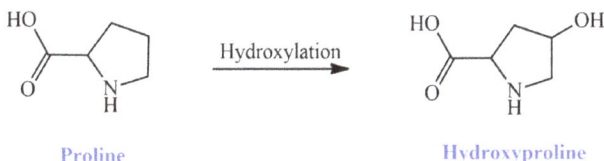

Figure 4.2: Synthesis of hydroxyproline.

Cysteine is oxidized under mild conditions to the disulfide cystine (Figure 4.3). The reaction is reversible. This linkage is important in maintaining the overall shape of a protein.

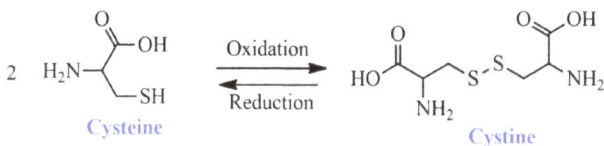

Figure 4.3: Oxidation of cysteine to cystine.

4.2.2 Synthesis of α-Amino Acids

4.2.2.1 Direct Ammonolysis of α-Halo Acid

Yields tend to be poor in this reaction. This is simply a nucleophilic substitution in which ammonia reacts with an α-halo carboxylic acid. An example is the synthesis of alanine from α-bromopropionic acid (Figure 4.4).

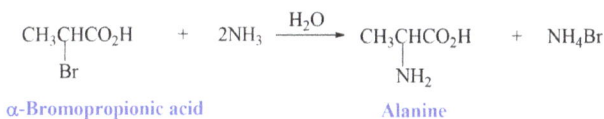

Figure 4.4: Synthesis of alanine from α-bromopropionic acid.

4.2.2.2 From Potassium Phthalimide

This is a variation of the Gabriel synthesis and yields are usually high. This is a good procedure to make glycine from potassium phthalimide (Figure 4.5).

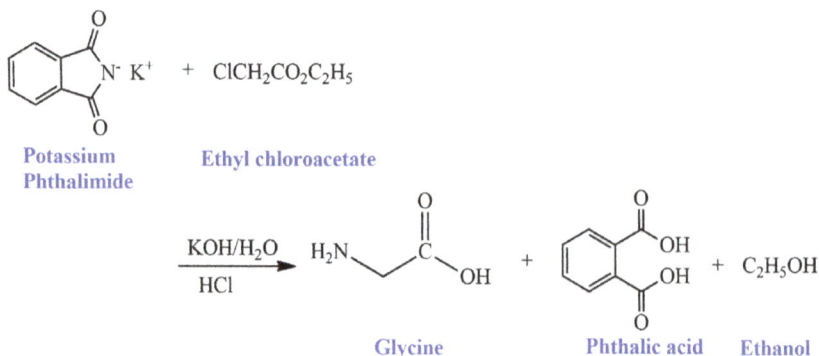

Potassium Phthalimide   **Ethyl chloroacetate**

Glycine   **Phthalic acid**   **Ethanol**

Figure 4.5: Synthesis of glycine from potassium phthalimide.

4.2.2.3 The Strecker Synthesis

Treatment of an aldehyde with ammonia and hydrogen cyanide yields an α-aminonitrile which is hydrolyzed to the α-amino acid. The reaction proceeds via an intermediate imine. This is a good route for the synthesis of alanine from acetaldehyde (Figure 4.6).

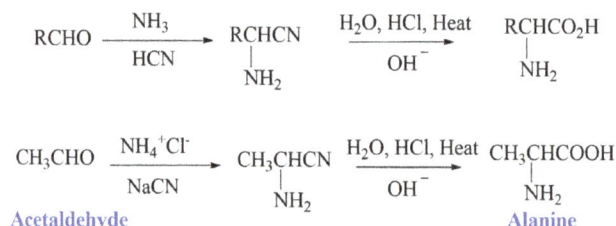

Acetaldehyde   **Alanine**

Figure 4.6: The Strecker synthesis.

4.2.3 Resolution of DL-Amino Acids: A Racemic Amino Acid Mixture

Conversion of DL-amino acids to a racemic mixture of N-acylamino acids and then hydrolysis with a deacylase enzyme that selectively deacylates the L-acylamino acid (Figure 4.7).

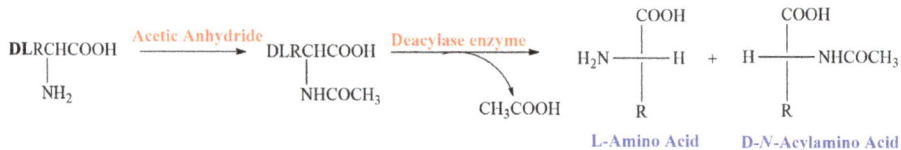

L-Amino Acid   D-N-Acylamino Acid

Figure 4.7: Resolution of DL-amino acids.

4.2.4 Acid–Base Behaviors of Amino Acids

In the dry solid state, amino acids exist as dipolar ions (zwitterions). In aqueous solution, equilibrium exists between the dipolar ions, that is, the cationic and the anionic forms of the amino acid. The predominant form depends on the pH of the solution. At low pH, the amino acid exists primarily in the cationic form and net charge = +1. At high pH, the amino acid exists primarily in the anionic form and net charge = −1. At some intermediate pH called the isoelectric point (*pI or IEP, is the pH at which amino acid carries no net electrical charge*), the concentration of the dipolar ion is at a maximum and the concentrations of anionic and cationic forms are equal and net charge = 0 (Figure 4.8).

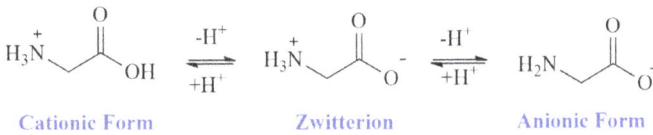

Figure 4.8: Acid–base behaviors of amino acids.

4.2.5 Separation of Amino Acids

Amino acids can be separated and purified by electrophoresis. This method depends on the movement of charged particles in an electric field. A mixture of amino acids is placed at the center of a sheet of cellulose acetate. The sheet is soaked with an aqueous buffered solution at a pH of 6.0. At this pH, aspartic acid will exist as its −1 ion, alanine as its zwitterions, and lysine as its +1 ion. Application of an electric current causes the negatively charged amino acid (aspartic acid) to migrate to the (+) electrode and the positively charged one (lysine) migrates to the (−) electrode. The amino acid (alanine) that exists as zwitterions, with a net charge of zero, remains at its original position (Figure 4.9).

4.2.6 Chirality of Amino Acids

In all chiral amino acids, the *stereogenic carbon* at the *chiral center* of the structure has four different groups attached to it. Amino acids can exist in two forms: L and D. In glycine, the "R" group is another hydrogen atom that is why glycine is achiral (Figure 4.10).

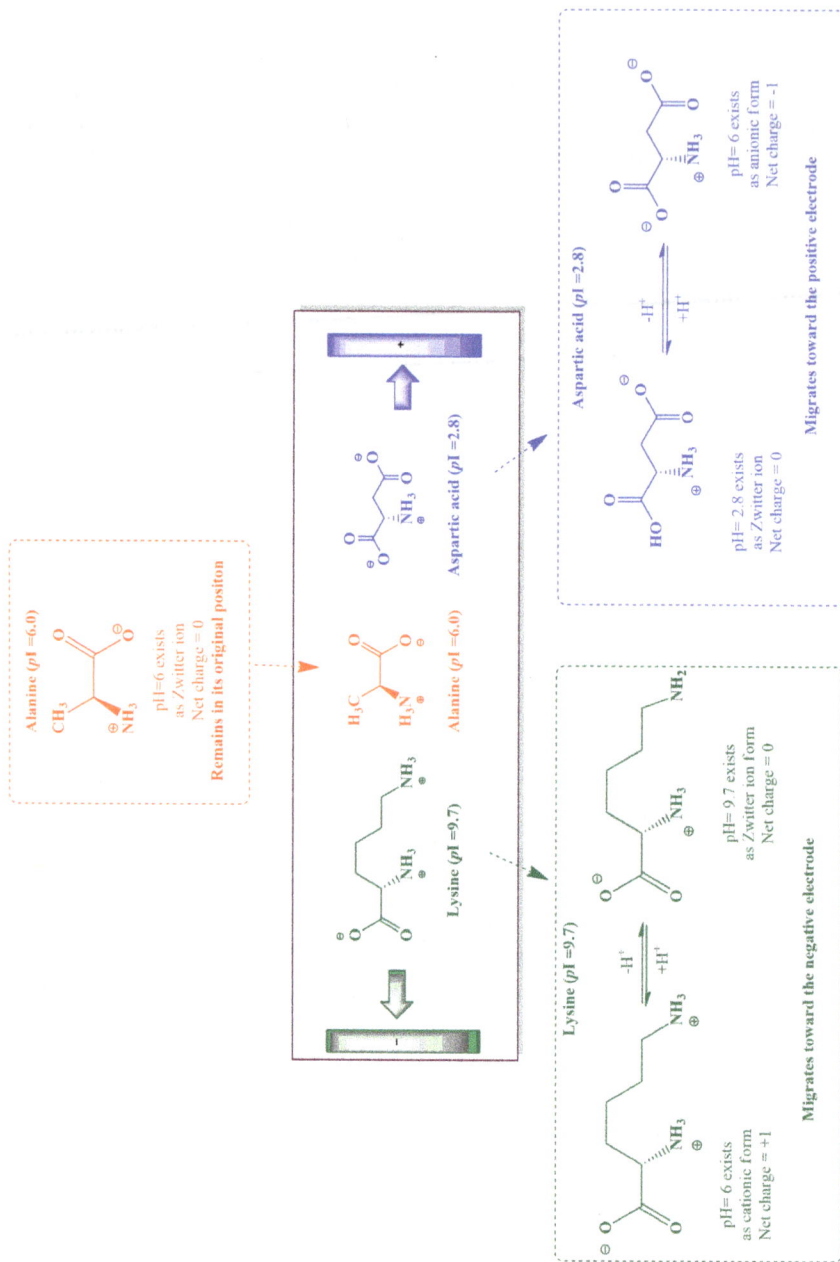

Figure 4.9: Separation of amino acids.

Four different groups
attached to this carbon
The central carbon is chiral or stereogenic

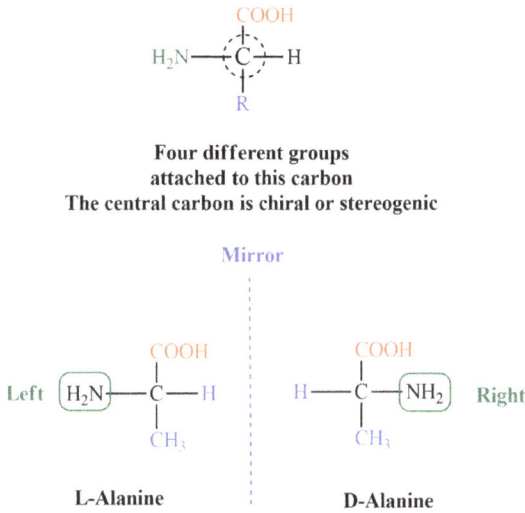

Figure 4.10: Chirality of amino acids.

4.2.7 Acid–Base Properties of Amino Acids

In the amino acid structure, there are both a basic amino group and an acidic carboxyl group. Amino acid can exist in the zwitterion form by the internal transfer of a hydrogen ion from the -COOH group to the -NH_2 group to leave an ion with both a negative charge and a positive charge (Figure 4.11).

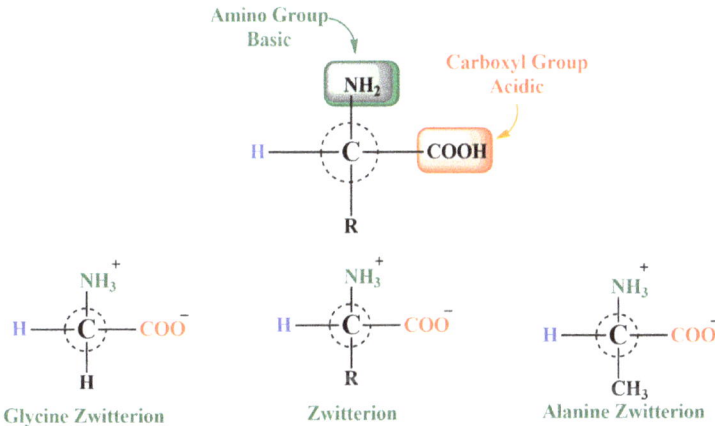

Figure 4.11: Acid–base properties of amino acids.

4.3 Peptide Bond

When -COOH group of one amino acid molecule reacts with -NH$_2$ group of another amino acid molecule, a molecule of water "H$_2$O" is released, and a peptide bond or amide bond "CO-NH" is formed (Figure 4.12). This type of reaction is called dehydration reaction (also known as *a condensation reaction*).

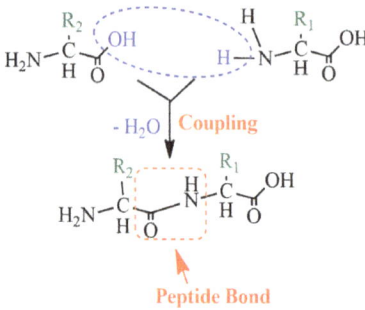

Figure 4.12: Peptide bond formation.

4.4 Polypeptides (Polypeptide Chains)

Amino acids join together to form unbranched amino acid chains called peptides, and when the number of amino acids in the chain is large then it is called a polypeptide chain (Figure 4.13). The length of each peptide chain varies from several amino acids to many amino acids. They are further classified by the number of amino acids present in the chain. A peptide having two amino acids is called a dipeptide, three amino acids is called a tripeptide, and so on. Number of Rs in the chain represents the number of amino acids, and the number of Rs − 1 represents the number of peptide bond in the same chain.

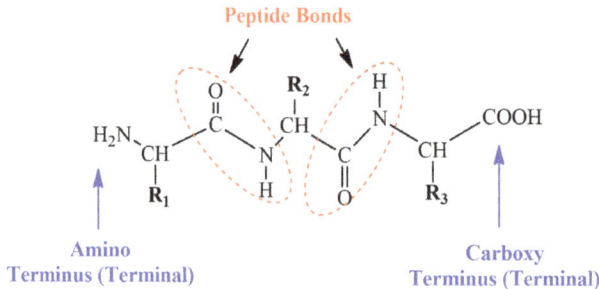

Figure 4.13: Polypeptide chain.

4.5 Peptide Nomenclature

The following IUPAC rules can be applied to name small peptides:

- Rule 1: The C-terminal amino acid residue (located at the far right of the structure) maintains its full amino acid name.
- Rule 2: All other amino acid residues have names that end in -yl. The -yl suffix replaces the -ine or -ic acid ending of the amino acid name, except for tryptophan (tryptophyl), cysteine (cysteinyl), glutamine (glutaminyl), and asparagine (asparaginyl).
- Rule 3: The amino acid naming sequence begins at the N-terminal amino acid residue.

For example, the IUPAC names of the following small peptides are:

Glu–Ser–Ala   Gly–Tyr–Gln–Asn   Gly–Tyr–Ser–Ser

In the first peptide chain, the three amino acids present are glutamic acid, serine, and alanine. Alanine, the C-terminal residue (on the far right), keeps its full name. The other amino acid residues in the peptide receive "shortened" names that end in -yl. The -yl replaces the -ine or -ic acid ending of the amino acid name. Thus, glutamic acid becomes glutamyl, serine becomes seryl, and alanine remains alanine. The IUPAC name, which lists the amino acids in the sequence from N-terminal residue to C-terminal residue, becomes glutamylserylalanine.

  In the second peptide chain, the four amino acids present are glycine, tryptophan, glutamine, and asparagine. The IUPAC name, which lists the amino acids in the sequence from N-terminal residue to C-terminal residue, becomes glycyltryptophylglutaminylasparagine.

  In the third peptide chain, the IUPAC name is glycyltyrosylserylserine.

4.6 Peptide Synthesis (The General Principles)

For the purpose of peptide synthesis, amino acids can be considered as having two main functionalities to manipulate, that is, the α-amino and carboxyl groups. Functional groups are also present in the side chains of many of amino acids. These functionalities must be protected so that they do not interfere with the formation of the peptide bond. With respect to the peptide bond formation, there are four main steps: protection- activation- coupling-, selective deprotection or deprotection as shown in Figure 4.14.

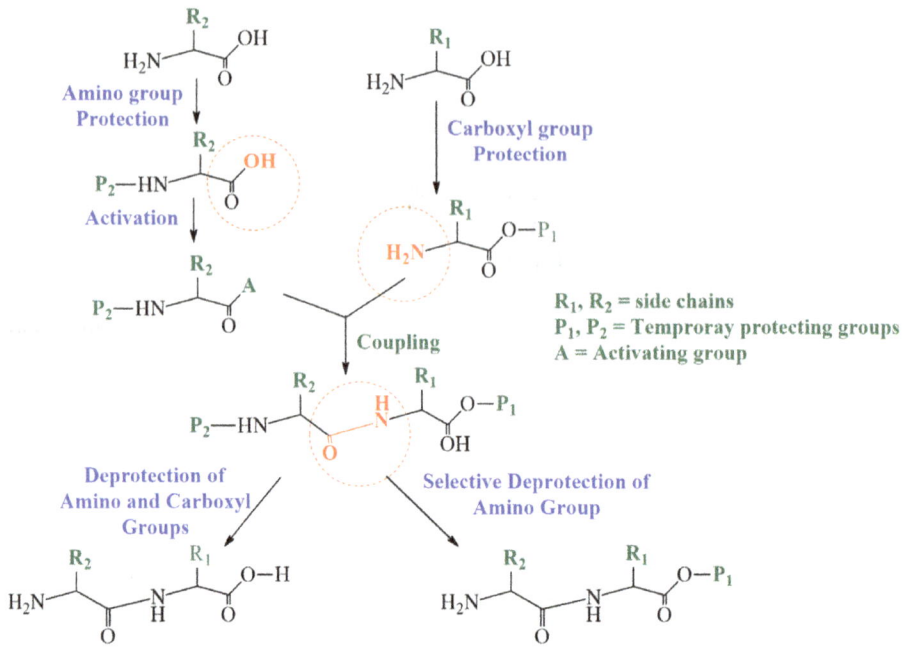

Figure 4.14: The general principles of the peptide synthesis.

4.6.1 Dipeptide Synthesis

Four steps are required to synthesize a dipeptide such as Ala–Gly as shown in Figure 4.15.

4.6.2 Solid-Phase Synthesis for Peptides and Polypeptides

Solid-phase peptide synthesis is an automated synthesis used for the production of synthetic peptides and polypeptides. The synthesis is based on the stepwise construction of a peptide chain attached to a solid support as described in Figure 4.16.

4.7 Enzymes

An enzyme is an organic compound that acts as a catalyst for a biochemical reaction. Enzymes catalyze and accelerate the reaction but are not consumed during the reaction. Some enzymes are simple proteins, consisting entirely of amino acid chains. Others are conjugated proteins containing additional chemical components.

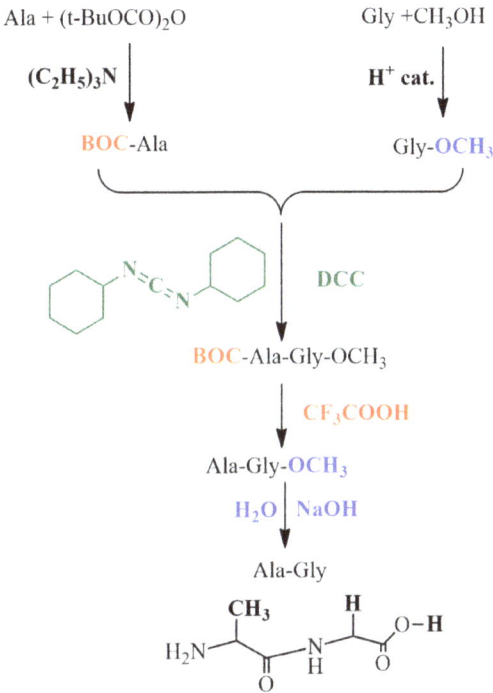

Figure 4.15: Dipeptide synthesis.

4.7.1 Enzyme Structure

Enzymes have two general structural classes: simple enzymes and conjugated enzymes. A simple enzyme is composed only of protein (amino acid chains). A conjugated enzyme has a nonprotein part called a cofactor in addition to a protein part called an apoenzyme. The apoenzyme and the cofactor are combined to form the holoenzyme which is the biochemically active enzyme.

4.7.2 Enzyme Nomenclature and Classification

The three important aspects of naming the enzyme are:
i.   The suffix -ase identifies the enzyme, for example, suc**rase** and lip**ase**.
ii. The prefix -ase is usually the type of the reaction catalyzed by an enzyme, for example, **oxidase** enzyme catalyzes an oxidation reaction, and a **hydrolase** enzyme catalyzes a hydrolysis reaction.
iii. Substrate identity is also added to the type of reaction, for example, **glucose** oxidase and **pyruvate** carboxylase.

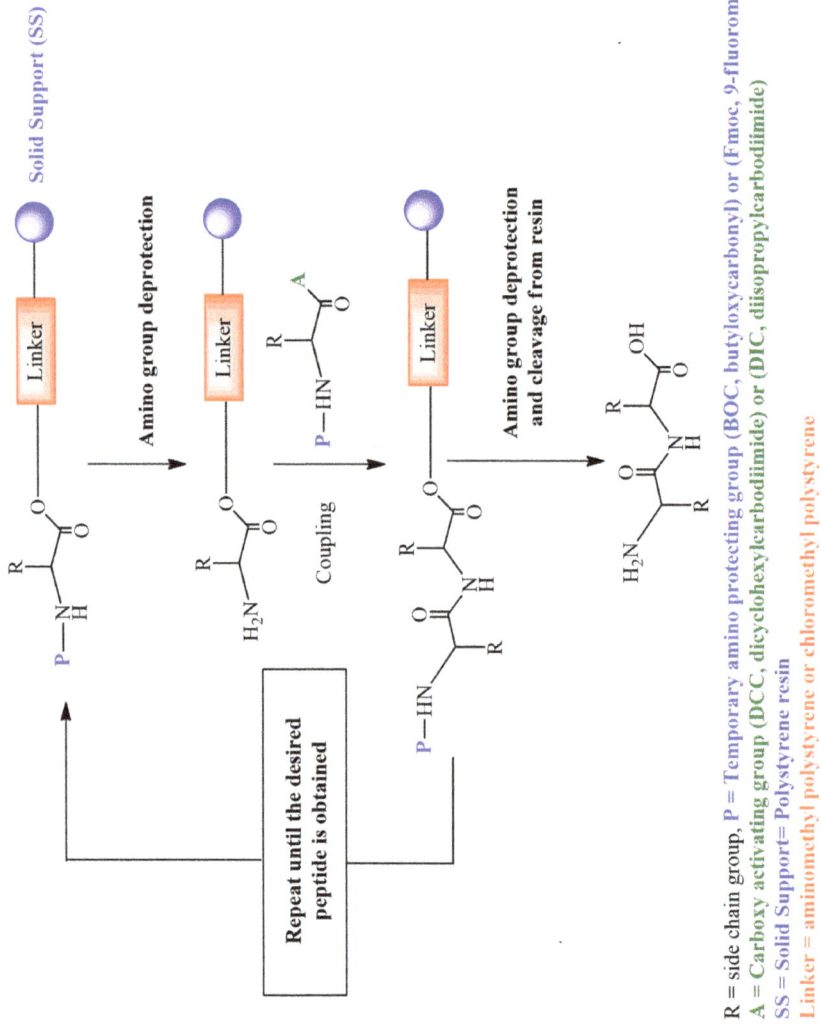

Figure 4.16: Solid-phase synthesis of peptides and polypeptides.

R = side chain group, P = Temporary amino protecting group (BOC, butyloxycarbonyl) or (Fmoc, 9-fluoromethyl carbamate)
A = Carboxy activating group (DCC, dicyclohexylcarbodiimide) or (DIC, diisopropylcarbodiimide)
SS = Solid Support= Polystyrene resin
Linker = aminomethyl polystyrene or chloromethyl polystyrene

Enzymes can be classified into six major classes based on the reactions they catalyze:

i.   An oxidoreductase catalyzes an oxidation–reduction reaction.
ii. A transferase catalyzes the transfer of a functional group from one molecule to another.
iii. A hydrolase catalyzes a hydrolysis reaction by the addition of a water molecule.
iv. A lyase catalyzes the addition of a group to a double bond or the removal of a group to form a double bond.
v.   An isomerase catalyzes the isomerization of a substrate in a reaction.
vi. A ligase catalyzes the bonding together of two molecules into one with the involvement of ATP.

4.7.3 Models of Enzyme Action

4.7.3.1 Lock-and-Key Model

In the lock-and-key model, the active site geometry is fixed and only substrates with a complementary geometry can fit in the site (Figure 4.17).

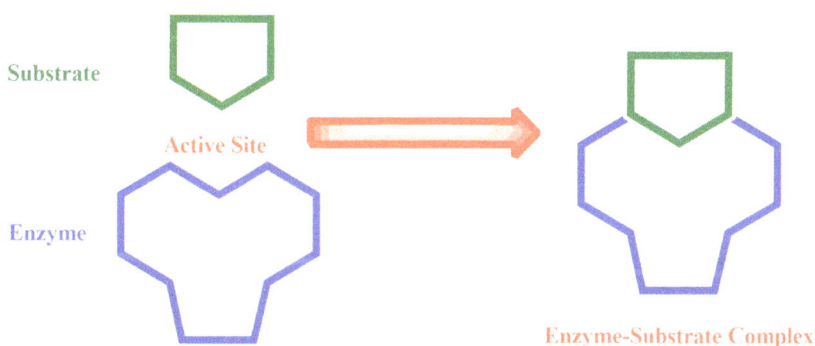

Figure 4.17: Lock-and-key model.

4.7.3.2 The Induced-Fit Model

This model allows minor changes in the geometry of the active site of the enzyme to accommodate the substrate (Figure 4.18).

4.7.4 Factors That Affect the Rate of Enzyme Activity

4.7.4.1 Temperature

The temperature increases the reaction rate until the protein is denatured, and activity falls distinctly.

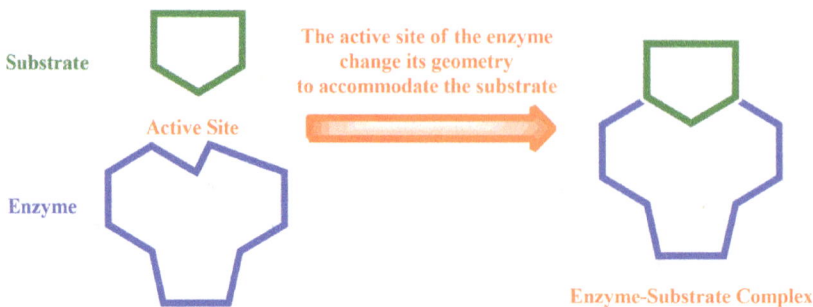

Figure 4.18: The induced-fit model.

4.7.4.2 pH
Maximum enzymatic activity can be kept within a narrow pH range; outside this range, the protein is denatured, and activity falls distinctly.

4.7.4.3 Concentration of Substrate
As the substrate concentration increases, the reaction rate increases until full saturation occurs, then the rate diminishes.

4.7.4.4 Concentration of Enzyme
Although the enzyme concentration is always much lower than the substrate concentration, increasing the enzyme concentration increases the reaction rate.

4.8 Summary and Important Concepts

Proteins and peptides. These are large biomolecules made of amino acid residues linked together by amide, or peptide, bonds.

Amino acids. These are organic compounds containing an amino group, and a carboxylic acid group attached to the carbon atom.

Standard amino acid. A standard amino acid is one of the 20 α-amino acids that are normally present in protein.

Amino acid classifications. Amino acids are classified as nonpolar, polar neutral, polar basic, or polar acidic depending on the nature of the side chain (R group) present in the structure.

Chirality of amino acids. Amino acids found in proteins are always left-handed (L isomer).

Zwitterion. A zwitterion is a molecule that has a positive charge on one atom and a negative charge on another atom.

Peptide bond. This is a chemical bond that is formed between two molecules of amino acids when the carboxyl group of one molecule reacts with the amino group of the other molecule, releasing a molecule of water.

Transamination. This is a reaction of moving an amino group from one molecule to another.

Isoelectric point (pI or IEP). This is the pH at which amino acid carries no net electrical charge.

Electrophoresis. A method depends on the movement of charged particles in an electric field used to separate and purify amino acids.

Solid-phase peptide synthesis. This is a major automated synthesis method or technology used for the production of synthetic peptides.

Enzymes. Enzymes are protein molecules that act as catalysts. Enzymes have names that provide information about their function. Most enzyme names have the suffix -ase.

Enzyme structure. Simple enzymes are made from amino acids protein only. Conjugated enzymes have a nonprotein portion (cofactor) in addition to a protein portion (the apoenzyme).

Enzyme classification. Enzymes are classified based on their function into oxidoreductases, transferases, hydrolases, lyases, isomerases, and ligases.

Enzyme active site. The enzyme active site is the small part in which the substrate binds to the enzyme.

Lock-and-key model. In this model, the active site of the enzyme has a fixed geometry. Only substrates with a complementary geometry can fit.

Induced-fit model. In this model, the active site can undergo small changes in geometry to accommodate the substrate.

Enzyme activity. Enzyme activity is a measure of the rate at which an enzyme converts the substrate to products. Four factors that affect enzyme activity are temperature, pH, substrate concentration, and enzyme concentration.

4.9 Practice Exercises

4.9.1 Why all amino acids (except glycine) are chiral?

4.9.2 Draw the chemical structure of leucine zwitterion.

4.9.3 Draw the general chemical structure of serine and show how a peptide bond is formed between two serine molecules.

4.9.4 List down the four main steps that are involved in the peptide synthesis.

4.9.5 Explain how alanine is prepared from 2-bromopropanoic acid.

4.9.6 What is the main difference between cationic and anionic forms of amino acids?

4.9.7 Draw chemical structures of alanine and tyrosine zwitterions.

4.9.8 Draw the general structure for L-amino acids.

4.9.9 Which one of the following amino acids is achiral?
i.   Lysine
ii. Alanine
iii. Glycine

4.9.10 Which of the following amino acids has its α-carbon as part of a five-membered ring?
i.   Proline
ii. Tryptophan
iii. Histidine

4.9.11 Which of the following amino acids has a nonpolar side chain?
i.   Serine
ii. Phenylalanine
iii. Asparagine

4.9.12 Provide the Fischer projection of L-serine.

4.9.13 Draw structures for the forms of glycine present in basic, neutral, and acidic solutions.

4.9.14 How many peptide bonds are there in the following peptide chain?

4.9.15 In which of the following the isoelectric point is important?
i.   Electrophoresis
ii. Determination of the C-terminal amino acid
iii. Determination of the N-terminal amino acid

4.9.16 How can an α-ketoacid be converted to an α-amino acid?

4.9.17 Explain why alanine is chiral and glycine is achiral.

4.9.18 Which indicator is commonly used to visualize bands of amino acids?

4.9.19 When a disulfide linkage is formed, the compound containing this new linkage has been _____.
i.   oxidized
ii. reduced
iii. hydrolyzed

4.9.20 The solid-phase method of peptide synthesis was devised by
i.   Sanger
ii. Merrifield
iii. Strecker

4.9.21 Which functional groups in the amino acid react to form the peptide bond?

4.9.22 What coupling reagent is commonly used in solid-phase peptide synthesis?

4.9.23 Show how peptide bond is formed between tyrosine and phenylalanine.

4.9.24 Predict the function of the following enzymes:
i.   Cellulase
ii. Sucrase
iii. L-Amino acid oxidase

4.9.25 Show how you would use a Strecker synthesis to make glycine.

Chapter 5
Nucleic Acids

5.1 Introduction

Nucleic acids are of two types: deoxyribonucleic acid (abbreviated as DNA) and ribonucleic acid (abbreviated as RNA). The monomeric units (the building blocks) of nucleic acids are called nucleotides (Figure 5.1).

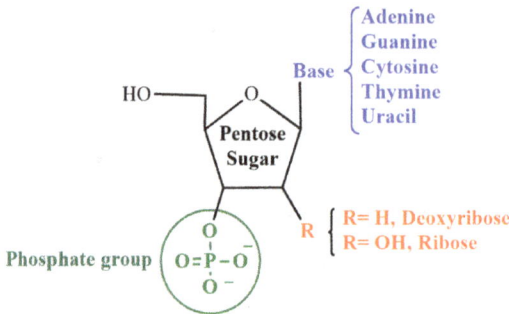

Figure 5.1: Nucleotide structure.

5.2 "Structure and Nomenclature" of Pentose Sugars

RNA and DNA differ in the sugar unit structure in their nucleotides. In RNA, the sugar unit is ribose (2-OH) and hence the R in RNA. In DNA, the sugar unit is 2-deoxyribose (2-H) and hence the D in DNA (Figure 5.2).

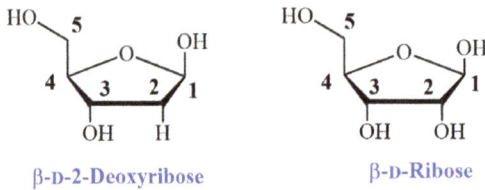

β-D-2-Deoxyribose   β-D-Ribose

Figure 5.2: Pentose sugars.

https://doi.org/10.1515/9783110762761-005

5.3 "Structure and Nomenclature" of Nucleic Acid Bases (Nitrogen-Containing Heterocyclic Bases)

There are five major bases (Figure 5.3). The derivatives of purine are called adenine and guanine, and the derivatives of pyrimidine are called thymine, cytosine, and uracil. The common abbreviations used for these five bases are, A, G, T, C, and U (uracil is thymine without methyl group). Cytosine, thymine, and uracil are numbered the same way. Numbering of guanine and adenine also follows the same system.

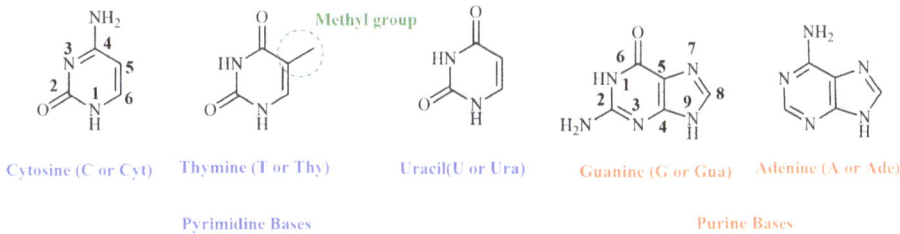

Cytosine (C or Cyt)   Thymine (T or Thy)   Uracil(U or Ura)   Guanine (G or Gua)   Adenine (A or Ade)

Pyrimidine Bases   Purine Bases

Figure 5.3: Nucleic acid bases.

5.4 "Structure and Nomenclature" of Nucleosides

Nucleosides are synthesized through the coupling of pyrimidine, purine, with ribose or deoxyribose sugars as graphically presented in Figure 5.4.

D-Ribose and 2′-deoxy-D-ribose are linked to purines and pyrimidines through the β-N-glycosidic bond between the anomeric carbons of the ribose and the 2′-deoxy-D-ribose and the N9 of the purines or N1 of the pyrimidines (Figure 5.5). Pentose ring atoms are designated with primed numbers. Nitrogen-containing base ring atoms are designated with unprimed numbers.

5.5 "Structure and Nomenclature" of Nucleotides

Nucleotides are the phosphate esters of nucleosides. Each RNA nucleotide monomer contains ribose, a phosphate group, and one of the heterocyclic bases adenine, cytosine, guanine, and uracil. DNA nucleotide monomer contains deoxyribose, a phosphate group, and one of the heterocyclic bases adenine, cytosine, guanine, and thymine (Figure 5.6). The phosphate group can be attached to 5′-OH or 3′-OH position.

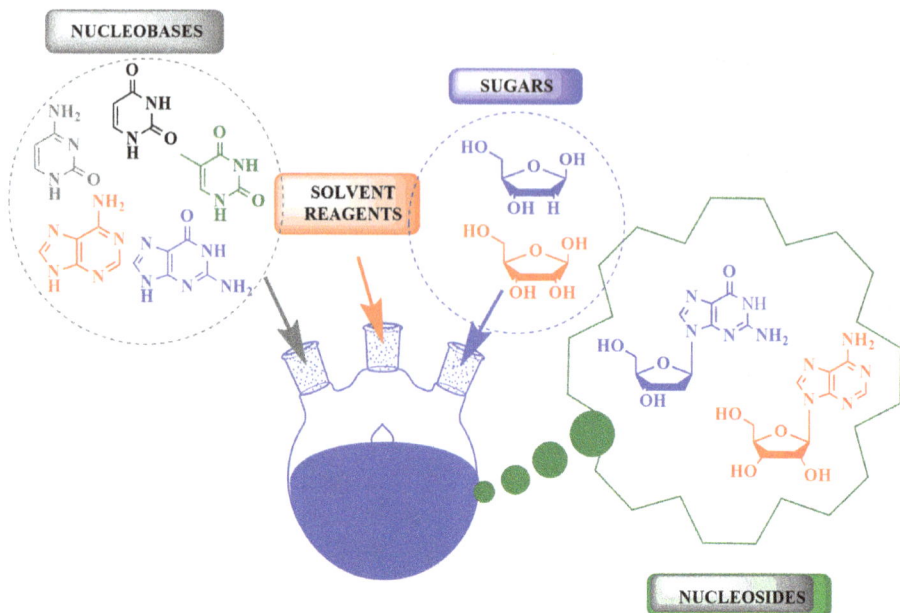

Figure 5.4: Graphical presentation of nucleoside synthesis.

5.6 Components of Nucleic Acids

Nucleic acids have the same components except for thymine and deoxyribose in DNA, and uracil and ribose in RNA as shown in Table 5.1 and Figure 5.7.

5.7 Comparison Between Nucleic Acids

There is a similarity between RNA and DNA: both are sugar-phosphate polymers, and both have nucleobases attached to the pentose sugars but there are few differences in structure and functions as shown in Table 5.2.

5.8 Syn Versus Anti-conformations

The nucleic acid base can exist in two distinct orientations about the *N*-glycosidic bond. These conformations are identified as *syn* and *anti*. It is the anti-conformation that predominates in nature (Figure 5.8).

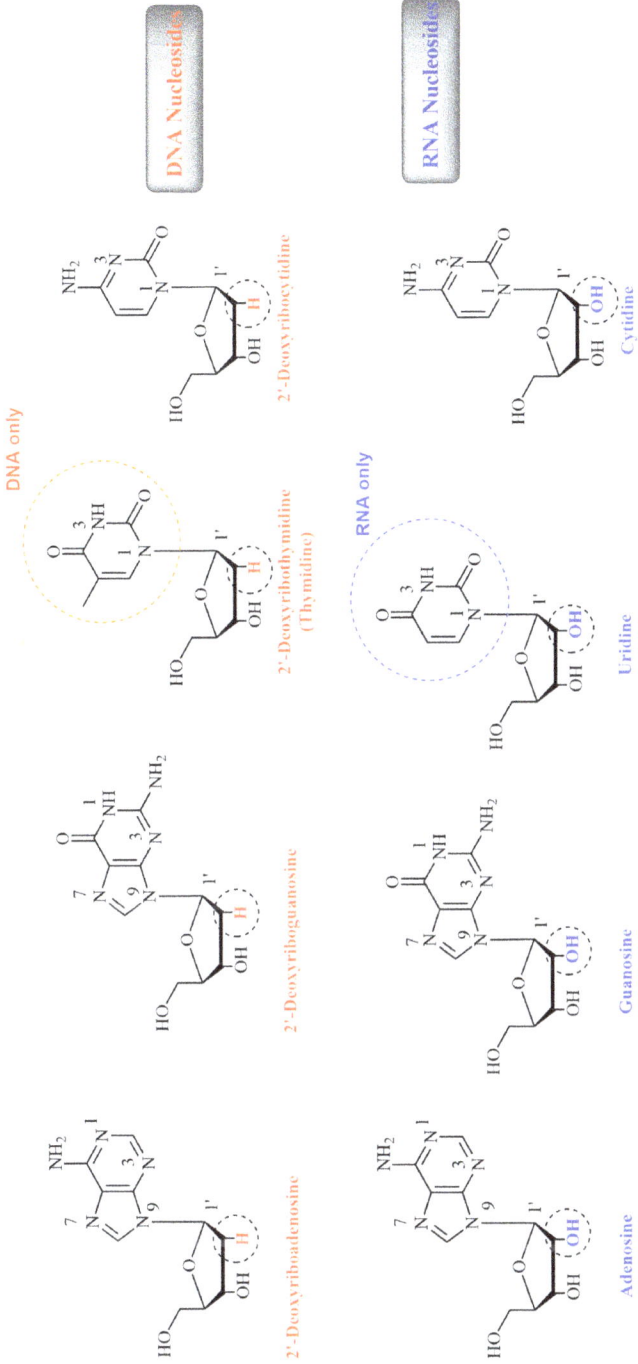

Figure 5.5: DNA and RNA nucleosides.

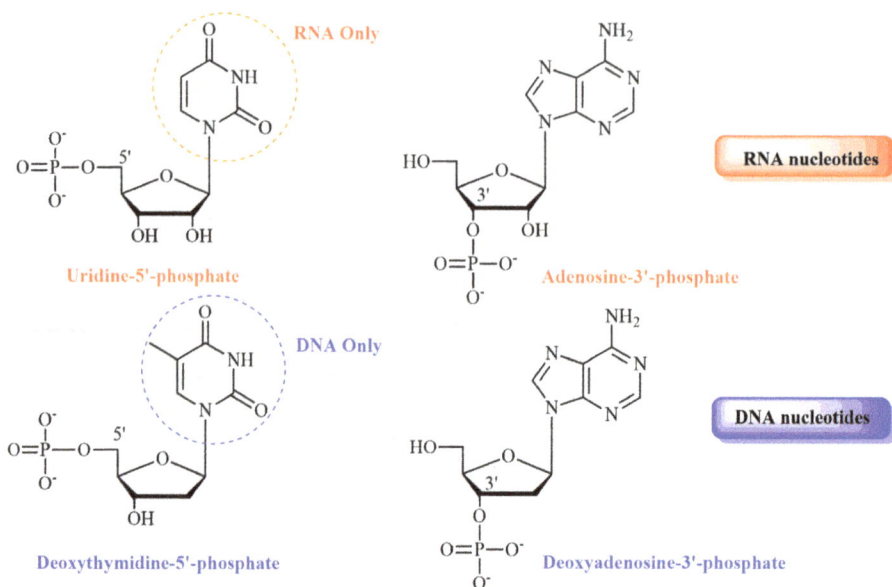

Figure 5.6: Examples of DNA and RNA nucleotides.

Table 5.1: Nucleic acid components.

DNA	Thymine	Cytosine	Guanine	Adenine	Phosphate	*Deoxyribose*
RNA	Uracil	Cytosine	Guanine	Adenine	Phosphate	Ribose

Thymine and deoxyribose are found in DNA only. Uracil and ribose are found in RNA only. Cytosine, adenine, guanine, and phosphate all exist in both DNA and RNA.

5.9 Tautomerism of the Nucleic Acid Bases

The nucleic acid bases (purines and pyrimidines) of the nucleotides exist in the hydroxyl pyrimidine or purine form that may add stabilization to the aromatic ring or in an amide-like structure. Despite the added stabilization by hydroxyl form, these bases prefer amide-like structures (Figure 5.9).

5.10 Hydrogen Bonding in Nucleic Acid Bases

Complementary hydrogen-bonded base pairs, as proposed by Watson and Crick, adenine with thymine or uracil and guanine with cytosine are shown in Figure 5.10.

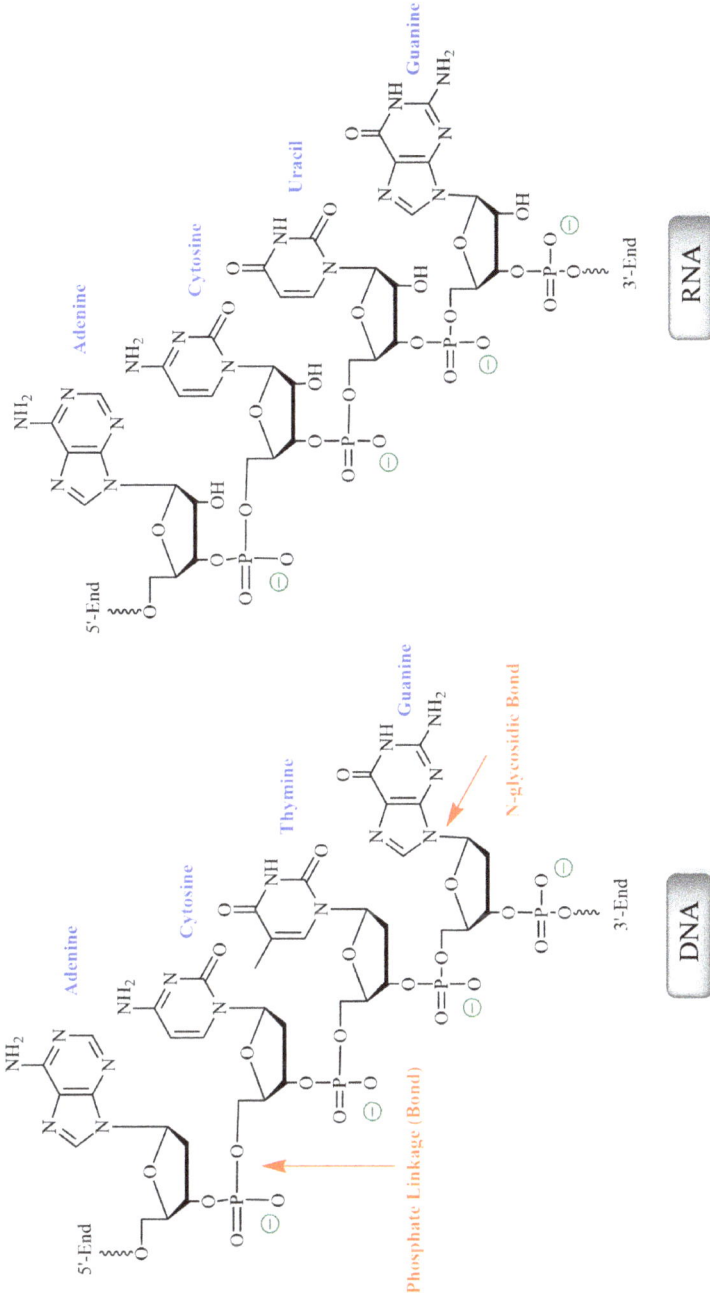

Figure 5.7: Nucleic acid components.

Table 5.2: Comparison of nucleic acids.

Nucleic acid	Sugar	Nucleobases	Shape	Function
DNA	Deoxyribose	Guanine Adenine Thymine Cytosine	Double-helix	Stores genetic material
RNA	Ribose	Guanine Adenine Uracil Cytosine	Single strand	**mRNA** encodes the copy of genetic information. **tRNA** carries amino acids for incorporation into protein. **rRNA** is a component of ribosomes.

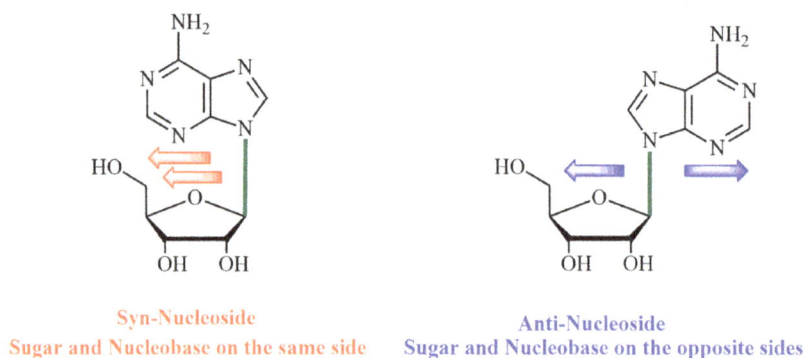

Syn-Nucleoside
Sugar and Nucleobase on the same side

Anti-Nucleoside
Sugar and Nucleobase on the opposite sides

Figure 5.8: Syn versus anti-conformations.

5.11 Solid-Phase Synthesis of Oligonucleotides

The building block used for this synthesis is the *phosphoramidite monomer* (Figure 5.11), in which protecting groups are added to amine, hydroxyl, and phosphate-reactive sites to prevent undesired side reactions and give rise to the formation of the correct final product. Once the synthesis is done, those protecting groups can be removed. The 3'-carbon is linked to the solid support, and synthesis proceeds from 3' to 5'. The solid support is made from a controlled pore glass beads with channels where the protected nucleotide is attached.

Oligonucleotide synthesis is done via a cycle of four chemical reactions that are repeated until the desired bases have been added as shown in Figure 5.12.

4-Amino-2-hydroxy-pyrimidine vs. cytosine   **2-Amino-6-hydroxypurine vs.** guanine

2,4-Dihydroxypyrimidine vs. uracil **(R=H)**
2,4-Dihydroxy-5-methyl-pyrimidine vs. -thymine **(R= CH₃)**

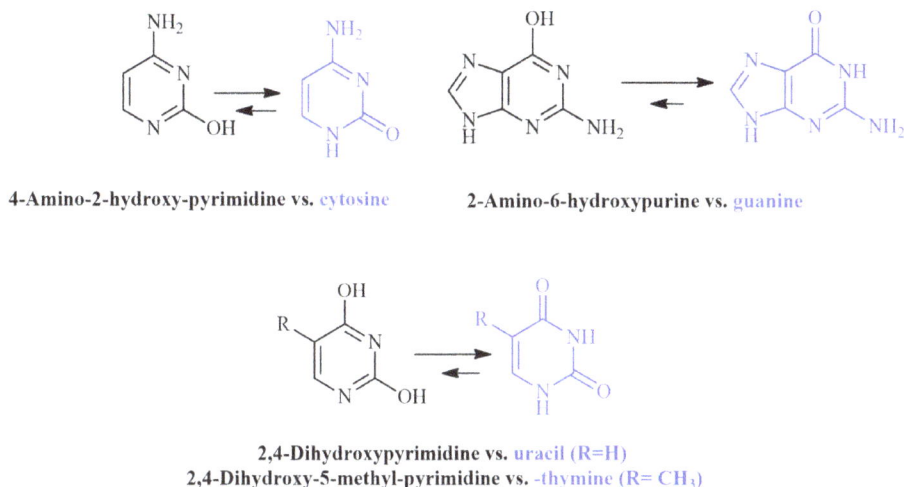

Figure 5.9: Tautomerism of the nucleic acid bases.

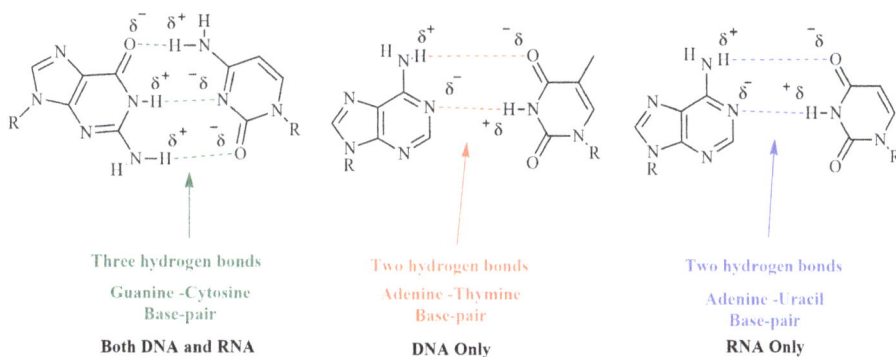

Three hydrogen bonds	Two hydrogen bonds	Two hydrogen bonds
Guanine -Cytosine Base-pair	Adenine -Thymine Base-pair	Adenine -Uracil Base-pair
Both DNA and RNA	**DNA Only**	**RNA Only**

Figure 5.10: Hydrogen bonding in nucleic acids bases.

5.12 Artificial Nucleic Acids (Modified Nucleic Acids)

There are different types of the modified nucleic acids. Among those are the peptide (or peptido) nucleic acid (PNA), morpholinos nucleic acid (MNA), glycol/glycerol nucleic acids (GNA), threose nucleic acid (TNA), and 4′-thionucleic acids (4′-SDNA).

5.12.1 Peptide (Peptido) Nucleic Acid (PNA)

The PNA differs in the chemical structure from DNA or RNA, where the sugar moiety is replaced by amino acids (Figure 5.13).

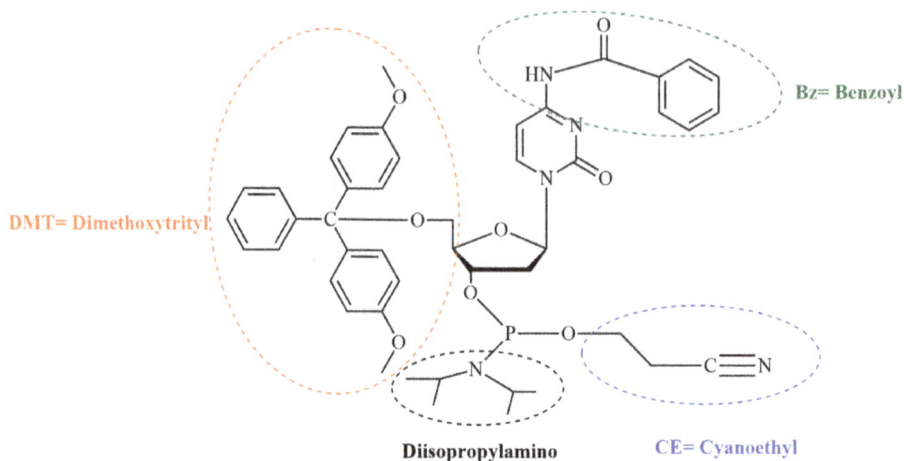

Figure 5.11: The phosphoramidite monomer.

5.12.2 Morpholino Nucleic Acid (MNA)

The main difference between MNA and DNA is that nucleic acid bases are bound to morpholine rings instead of deoxyribose sugars and linked through phosphorodiamidate groups instead of phosphates (Figure 5.14).

5.12.3 Glycerol–Glycol Nucleic Acid (GNA)

GNA (or glycol nucleic acid) differs from DNA and RNA by having repeating glycerol (three carbon atoms) or glycol (two carbon atoms) units linked by phosphodiester bonds (Figure 5.15).

5.12.4 Threose Nucleic Acid (TNA)

The structure of TNA is similar to DNA or RNA but differing in the composition of its "backbone." DNA and RNA have a deoxyribose and ribose sugar backbone, respectively, whereas the backbone of TNA composed of repeating threose units is linked together by phosphodiester bonds (Figure 5.16).

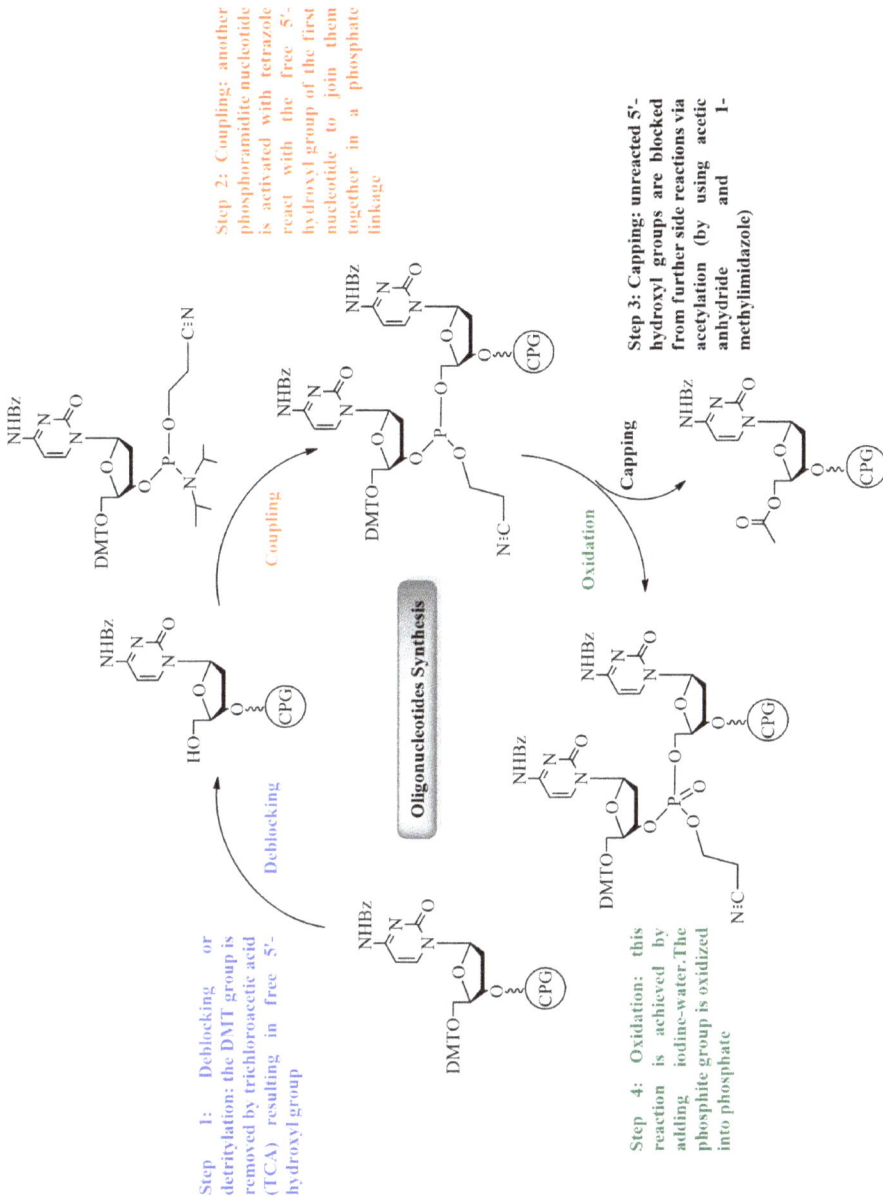

Step 2: Coupling: another phosphoramidite nucleotide is activated with tetrazole react with the free 5'-hydroxyl group of the first nucleotide to join them together in a phosphate linkage

Step 3: Capping: unreacted 5'-hydroxyl groups are blocked from further side reactions via acetylation (by using acetic anhydride and 1-methylimidazole)

Step 1: Deblocking or detritylation: the DMT group is removed by trichloroacetic acid (TCA) resulting in free 5'-hydroxyl group

Step 4: Oxidation: this reaction is achieved by adding iodine-water. The phosphite group is oxidized into phosphate

Figure 5.12: Oligonucleotides' solid-phase synthesis.

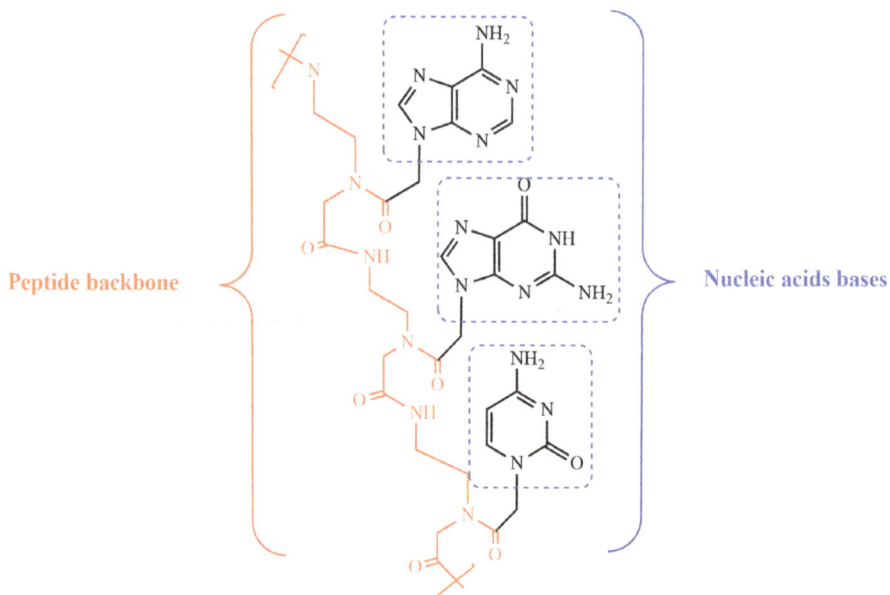

Figure 5.13: Peptide (peptido) nucleic acid.

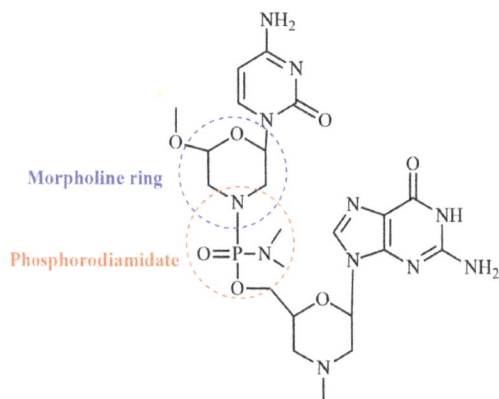

Figure 5.14: Morpholino nucleic acid.

5.12.5 4′-Thionucleic Acid (4′-ThioDNA or 4′-SDNA)

4′-SDNA has a structure similar to DNA or RNA but differs in the composition of its sugar moiety. 4′-SDNA backbone is composed of repeating 4′-thio-2′-deoxyribose units linked together by phosphodiester bonds (Figure 5.17).

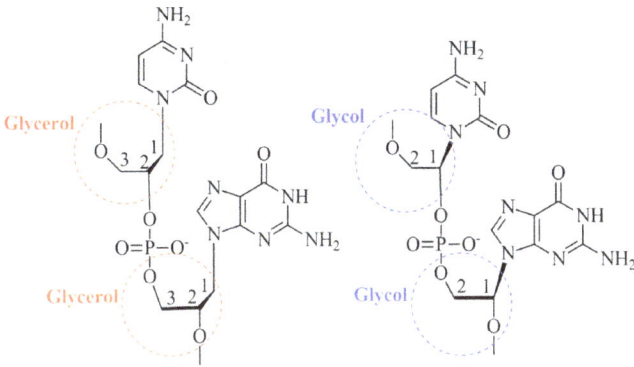

Figure 5.15: Glycerol and glycol nucleic acid.

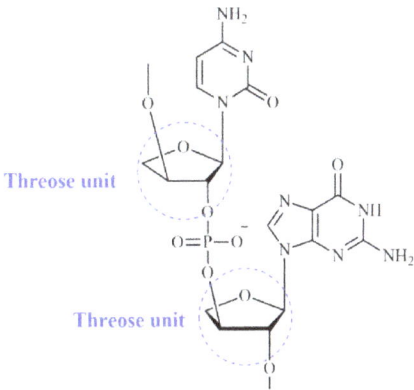

Figure 5.16: Threose nucleic acid.

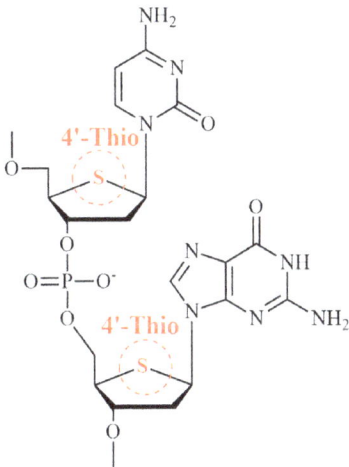

Figure 5.17: 4′-Thionucleic acid.

5.13 Chemistry of Heredity

5.13.1 The Genetic Code

The genetic code 64 triplets (codons) of nucleotides main rule is to encode amino acids. In the genetic code, each codon encodes for 1 of the 20 amino acids involved in the protein synthesis. There are RNA codons and DNA codons. RNA codons are the ones found in the messenger RNA and read during the translation process to produce the desired amino acid sequence.

5.13.1.1 The RNA Codons

In the following RNA codons, the left-hand column gives the first nucleotide of the codon, the four middle columns give the second nucleotide, and the last column gives the third nucleotide.

First nucleotide	Second nucleotide				Third nucleotide
	U	C	A	G	
U	UUU phenylalanine (Phe)	UCU serine (Ser)	UAU tyrosine (Tyr)	UGU cysteine (Cys)	U
	UUC (Phe)	UCC (Ser)	UAC (Tyr)	UGC (Cys)	C
	UUA leucine (Leu)	UCA (Ser)	UAA STOP	UGA STOP	A
	UUG (Leu)	UCG (Ser)	UAG STOP	UGG tryptophan (Trp)	G
C	CUU leucine (Leu)	CCU proline (Pro)	CAU histidine (His)	CGU arginine (Arg)	U
	CUC (Leu)	CCC (Pro)	CAC (His)	CGC (Arg)	C
	CUA (Leu)	CCA (Pro)	CAA glutamine (Gln)	CGA (Arg)	A
	CUG (Leu)	CCG (Pro)	CAG (Gln)	CGG (Arg)	G

(continued)

	AUU isoleucine (Ile)	ACU threonine (Thr)	AAU asparagine (Asn)	AGU serine (Ser)	U
A	AUC (Ile)	ACC (Thr)	AAC (Asn)	AGC (Ser)	C
	AUA (Ile)	ACA (Thr)	AAA lysine (Lys)	AGA arginine (Arg)	A
	AUG methionine (Met) or **START**	ACG (Thr)	AAG (Lys)	AGG (Arg)	G
	GUU valine (Val)	GCU alanine (Ala)	GAU aspartic acid (Asp)	GGU glycine (Gly)	U
G	GUC (Val)	GCC (Ala)	GAC (Asp)	GGC (Gly)	C
	GUA (Val)	GCA (Ala)	GAA glutamic acid (Glu)	GGA (Gly)	A
	GUG (Val)	GCG (Ala)	GAG (Glu)	GGG (Gly)	G

5.13.1.2 The DNA Codons

These DNA codons are read on the sense of 5′–3′ strand of DNA. Except the nucleotide thymidine (T) that is found in place of uridine (U), they read the same as RNA codons.

TTT	Phe	TCT	Ser	TAT	Tyr	TGT	Cys
TTC	Phe	TCC	Ser	TAC	Tyr	TGC	Cys
TTA	Leu	TCA	Ser	TAA	**STOP**	TGA	**STOP**
TTG	Leu	TCG	Ser	TAG	**STOP**	TGG	Trp
CTT	Leu	CCT	Pro	CAT	His	CGT	Arg
CTC	Leu	CCC	Pro	CAC	His	CGC	Arg
CTA	Leu	CCA	Pro	CAA	Gln	CGA	Arg
CTG	Leu	CCG	Pro	CAG	Gln	CGG	Arg
ATT	Ile	ACT	Thr	AAT	Asn	AGT	Ser
ATC	Ile	ACC	Thr	AAC	Asn	AGC	Ser
ATA	Ile	ACA	Thr	AAA	Lys	AGA	Arg
ATG	Met	ACG	Thr	AAG	Lys	AGG	Arg
GTT	Val	GCT	Ala	GAT	Asp	GGT	Gly

(continued)

GTC	Val	GCC	Ala	GAC	Asp	GGC	Gly
GTA	Val	GCA	Ala	GAA	Glu	GGA	Gly
GTG	Val	GCG	Ala	GAG	Glu	GGG	Gly

The following example shows the relationship between the DNA informational and template strand segments along with the protein segment for which the codes are illustrated.

Example 1

DNA informational strand	5* AUG CCA GUA GGC CAC UUG UCA 3*
DNA template strand	3* TAC GGT CAT CCG GTG AAC AGT 5*
mRNA	5* ATG CCA GTA GGC CAC TTG TCA 3*
Protein	Met Pro Val Gly His Leu Ser

Example 2

DNA informational strand	5* AAC GUU CAA ACU GUC 3*
DNA template strand	3* TTG CAA GTT TGA CAG 5*
mRNA	5* AAC GTT CAA ACT GTC 3*
Protein	Asn Val Gln Thr Val

5.14 Polymerase Chain Reaction (PCR)

The polymerase chain reaction (PCR) is usually used when a larger amount is required from an available tiny amount of DNA. The main purpose of PCR is to produce multiple copies of a given DNA sequence. For example, from 1 pg of DNA, it is possible to get several micrograms in few hours. The PCR uses Taq DNA polymerase which can construct the entire complementary strand from a single strand of DNA having a short, primer segment of complementary chain at one end. The overall process is summarized in the following three steps and is shown in Figure 5.18.

During the process, the tiny amount of the DNA to be amplified is heated in the presence of *Taq* polymerase, Mg^{2+} ion, the four deoxynucleotide triphosphate monomers, and a large excess of two short oligonucleotide primers of about 20 bases each. Each primer is complementary to the sequence at the end of one of the target DNA segments.

STEP 1

Denaturing: In this step, the temperature is raised to 95 °C so that the double-stranded DNA denatures into two single strands.

STEP 2

Annealing: When the DNA is denatured into two single strands the temperature is then lowered to 50 °C to force the DNA primers to attach to the template DNA by hydrogen bonding to their complementary sequence.

STEP 3

Extending: When the temperature is raised to 72 °C, a new strand of DNA is then made by the Taq polymerase and when replication of each strand is complete, two copies of the original DNA are produced.

Repeating the denature–anneal–extend cycle, a second time yields four DNA copies, repeating a third time yields eight copies, and so on, in an exponential series.

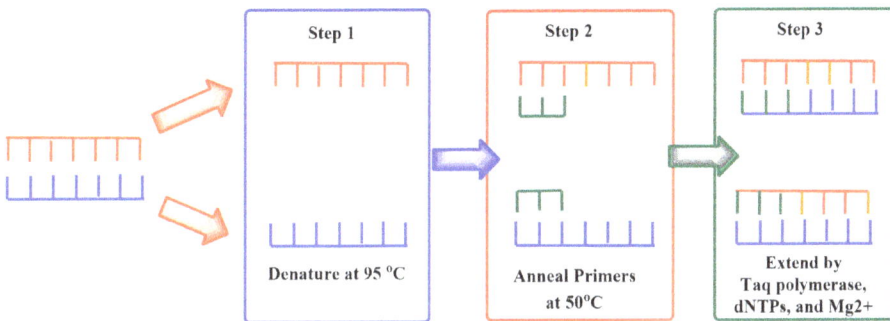

Figure 5.18: PCR overall process.

5.15 DNA Fingerprinting

The development of DNA fingerprinting arose from the finding that human genes contain short, repeating sequences of noncoding DNA, called short tandem repeat (STR) loci. The STR loci are slightly different for everyone except identical twins.

In addition to being used by crime laboratories to link suspects to biological evidence – blood, hair follicles, skin, or semen – found at a crime scene, DNA fingerprinting is also widely used for the diagnosis of genetic disorders, both prenatally and in newborns, cystic fibrosis, hemophilia, Huntington's disease, Tay–Sachs disease, sickle cell anemia, and thalassemia.

For use in criminal cases, the most accurate 13 core STR loci are adopted for the identification of any tested individual. Based on the 13 loci, a Combined DNA

Index System is put in place to serve as a registry of any convicted offenders. When a DNA sample is collected from a crime scene, the sample is subjected to cleavage with restriction endonucleases to cut out fragments containing the STR loci, the fragments are then amplified using PCR, and the sequences of the fragments are determined.

If the profile of sequences from a known individual and the profile sequence from DNA obtained at a crime scene match, the probability is approximately 82 billion to 1 that the DNA is from the same individual.

5.16 Summary and Important Concepts

Nucleic acids. They are polymers of nucleotides. Each nucleotide has a sugar, a nucleobase, and a phosphate group. The sugar is D-ribose in RNAs and 2-deoxy-D-ribose in DNAs.

DNA. The DNA consists of two polynucleotide strands twisted together in a double helix.

RNA. The RNA consists of one polynucleotide strand.

Nucleosides. Nucleosides are purine and pyrimidine bases linked to D-ribose or 2'-deoxy-D-ribose.

Nucleotides. Nucleotides are the phosphate esters of nucleosides.

Syn versus anti-conformations. The orientations of nucleic acid bases are around the N-glycosidic bond.

Oligonucleotide solid-phase synthesis. This is the synthesis of sequence of nucleotides by automated solid-phase techniques, where the nucleotide chain is built up by adding a protected phosphoramidite monomer to a protected nucleotide linked to a solid phase in the presence of a coupling agent.

Artificial nucleic acids. Modified nucleic acids are distinguished from naturally occurring DNA or RNA based on the changes to the backbone of the molecule.

Peptide nucleic acid. A chemical like DNA or RNA differs in the composition of its backbone.

Morpholino nucleic acid. Modified nucleic acids have standard nucleic acid bases bound to morpholine rings instead of deoxyribose rings and linked through phosphorodiamidate groups instead of phosphates.

Glycerol nucleic acid or glycol nucleic acid. A chemical like DNA or RNA differs in the composition of its backbone. GNA's backbone is composed of repeating glycerol (three carbon atoms) or glycol (two carbon atoms) units linked by phosphodiester bonds.

Threose nucleic acid. This is a chemical like DNA or RNA but differs in the composition of its backbone. TNA's backbone is composed of repeating threose units linked by phosphodiester bonds.

4'-Thionucleic acid. This is a chemical like DNA or RNA but differs in the composition of its sugar structure. 4'-SDNA backbone is composed of repeating 4'-thio-2'-deoxyribose units linked by phosphodiester bonds.

Polymerase chain reaction. This is a technique used when a larger amount is required from available tiny amount of DNA. PCR produces multiple copies of a given DNA sequence.

DNA fingerprinting. This is a technique used by crime laboratories to link suspects to biological evidence – blood, hair follicles, skin, or semen – found at a crime scene. DNA fingerprinting is also widely used for the diagnosis of genetic disorders, both prenatally and in newborns, cystic fibrosis, hemophilia, Huntington's disease, Tay–Sachs disease, sickle cell anemia, and thalassemia.

5.17 Practice Exercises

5.17.1 Name the protecting groups in the following phosphoramidite monomer:

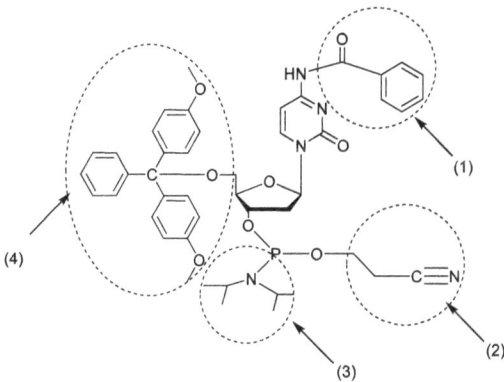

5.17.2 Write down the full names of the following artificial nucleic acids:
i. PNA
ii. TNA
iii. MNA

5.17.3 What DNA and RNA base sequences are complementary to the following DNA base sequence C-A-C-G-C-T-A-T-G-A-T-A-T-C-G-C-C-G?

5.17.4 List down the three main components of the nucleotides.

5.17.5 How many complementary hydrogen bonds occur when guanine and cytosine form a base pair?

5.17.6 Identify the conformation of the following nucleosides:

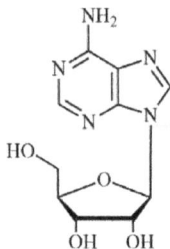

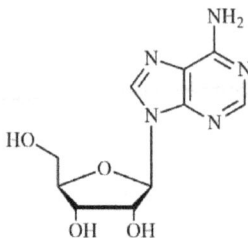

5.17.7 What are the first two steps in the oligonucleotide synthesis?

5.17.8 Assign numbers to the atoms in the structures of the following nucleic acid bases.

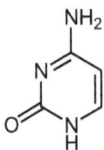

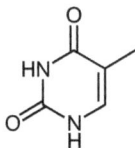

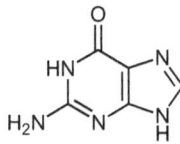

5.17.9 Give two differences between DNA and RNA components.

DNA	RNA

5.17.10 Besides a possible difference in base structure, what is the major structural difference between ribonucleosides and deoxyribonucleosides?

5.17.11 Write down the full name of the nucleic acid base G.

5.17.12 What is the major structural difference between a ribonucleoside and a ribonucleotide?

5.17.13 Which of the following nitrogens of adenine connects to ribose or deoxyribose to form a nucleoside?

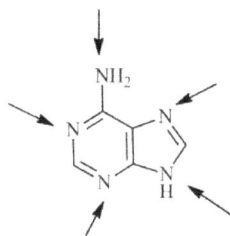

5.17.14 Which of the following nitrogens of thymine connects to deoxyribose to form a nucleoside?

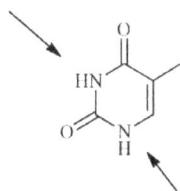

5.17.15 What are the four common ribonucleosides?

5.17.16 Name two pyrimidine bases that can exist in deoxyribonucleotides.

5.17.17 Show the hydrogen bonding that occurs when guanine and cytosine form a base pair.

5.17.18 Show the hydrogen bonding that occurs when adenine and thymine form a base pair.

5.17.19 Identify the following bases, and tell whether each is found in DNA, RNA, or both.

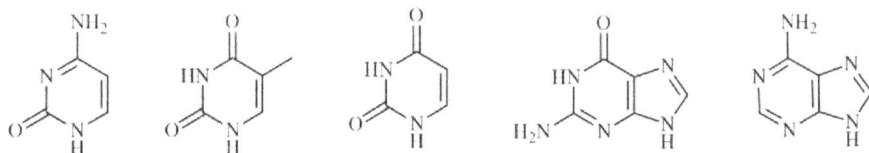

5.17.20 In this exercise, discover the consequences of two different kinds of mutations, **base substitution** ("BS") and **frame-shift** (insertion "Ins" or deletion "Del"), and practice using the genetic code to deduce amino acid sequences coded in four DNA (18 mer each) sequences. By doing the following:
- Deduce the RNA sequence resulting from transcription by using the knowledge of nucleic acids base pairing.
- From the information gained in step 1, deduce the amino acid sequence resulting from translation. Use tables in Sections 5.12.1.1 and 5.12.1.2.

- Describe the types of mutations for changing DNA 1 to DNA 2, DNA 1 to DNA 3, and DNA 2 to DNA 4.
- Explain why there was no change in the amino acid sequence even after six base substitutions, and why the single base difference between DNA 1 and 3 had such a big effect.

The four DNA sequences:

T-A-C-G-C-T-T-T-G-A-T-A-T-C-G-C-C-T
T-A-C-G-C-C^{BS}-T-T-A^{BS}-A-T-G^{BS}-A^{BS}-G^{BS}-G-C-C-C^{BS}
T-A-C-A^{Ins}-G-C-T-T-T-G-A-T-A-T-C-G-C-C^{Del}
T-A-C-T^{Ins}-G-C-C-T-T-A-A-T-G-A-G-G-C-C^{Del}

5.17.21 What amino acid sequence is coded for by the mRNA base sequence CAG-AUG-CCU-UGG-CCC-UUA?

5.17.22 Combine the structures below to create a ribonucleotide and draw the resulting ribonucleotide structure, and give it a name.

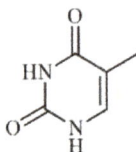

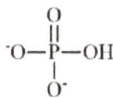

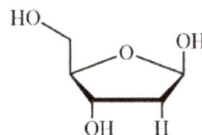

5.17.23 What are the sugars in DNA and RNA, and how do they differ?

5.17.24 For the following nucleotide structures, label the three nucleic acid building blocks they contain and draw a circle around the nucleoside portion of the molecule.

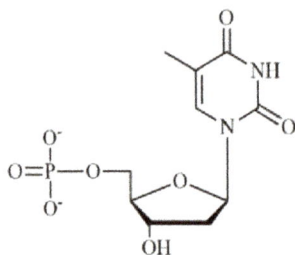

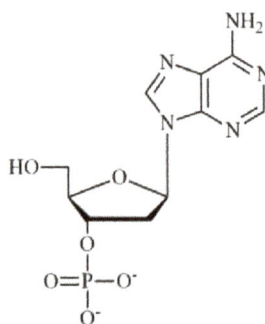

5.17.25 What are the three main types of RNA, and what are their functions?

Part II: **Synthetic Polymers**

Chapter 6
Industrial Synthetic Polymers

6.1 Definition of Polymer

Simply, a polymer may be defined as a substance built up of a number of repeating chemical units held together by chemical bonds (usually covalent bonds). Polymers differ from one another through the chemical and physical nature of the repeating units (RU) in the chains.

6.2 Classification of Polymers

6.2.1 Polymer Materials (According to Their Nature or Source of Origin) Can Be Classified into Three Major Types

- Natural organic polymers (biopolymers) are nucleic acids like DNA and RNA, polysaccharides like starch and cellulose, and proteins like polyglycine.
- Synthetic organic polymers are polyethylene (PE), polystyrene (PS), and polypropylene (PP).
- Semisynthetic organic polymers are vulcanized rubber and cellulose acetate.

6.2.2 Polymers (According to the Structure) Can Be Broadly Classified

- Linear polymers: These polymers are similar in structure and form long straight chains of identical links connected together. A common example is PVC (polyvinyl chloride) which is used for making electric cables and pipes.
- Branched chain polymers: The structure of these polymers is like branches originating at random points from the single linear chain. Low-density PE (LDPE) is used in plastic bags, and general-purpose container is a common example.
- Cross-linked or network polymers: Here the monomers are linked together to form a three-dimensional (3D) structure or network. These polymers are brittle and hard, for examples, bakelite and melamine.

6.2.3 Polymers Can also Be Classified (According to the Mode of Polymerization)

- Addition polymers can be formed by the addition of monomers to the growing chain (usually unsaturated molecules with double or triple bonds). PE and PP are examples of those polymers.

https://doi.org/10.1515/9783110762761-006

- Condensation polymers can be formed by the combination of two bifunctional monomers and releasing of small molecules. Nylon-6, 6 is an example.

6.2.4 Polymers Are also Classified (Based on Molecular Forces That Hold Atoms Together Within a Molecule) as Elastomers, Thermoplastics, Thermosetting, and Fibers

- Elastomers are rubber-like solid polymers that are elastic in nature. The most common examples are hair bands and rubber bands.
- Thermoplastic polymers are long-chain polymers when heated are softened (thick fluid-like) and hardened when cooled down. A common example is a PS.
- Fibers are thread-like in nature and can easily be woven. A common example is nylon-6, 6.
- Thermosetting polymers are semifluid in nature with low molecular masses. They form 3D structures on the application of heat. Bakelite is a common example.

6.3 Polymerization

Polymerization is a process by which small molecules (monomers) are converted to large molecules (polymers). This can be done as follows:

- Condensation polymerization is also described as a step polymerization or step-growth polymerization or step reaction polymerization and sometimes polycondensation.
- Addition polymerization is also called chain polymerization or chain-growth polymerization or chain reaction polymerization.

6.3.1 Condensation Polymerization

In condensation polymerization, monomers must have bifunctionalities. Each monomer needs to have at least two reactive sites (functional groups). Monomers that have only one functional group like ethanol and ethanoic acid cannot be used for condensation polymerization because the reaction cannot go further (Figure 6.1).

However, if each reacting molecule has two functional groups, then the reaction can proceed further. A good example is a reaction between the two monomers: hexanedioic acid (adipic acid) and ethane-1, 2-diol (ethylene glycol). The two bifunctional monomers have reacted together to make what is called a dimer. It can be seen that in this case the product still contains two functional groups, so further reaction with monomer can now take place to form a trimer (Figure 6.2).

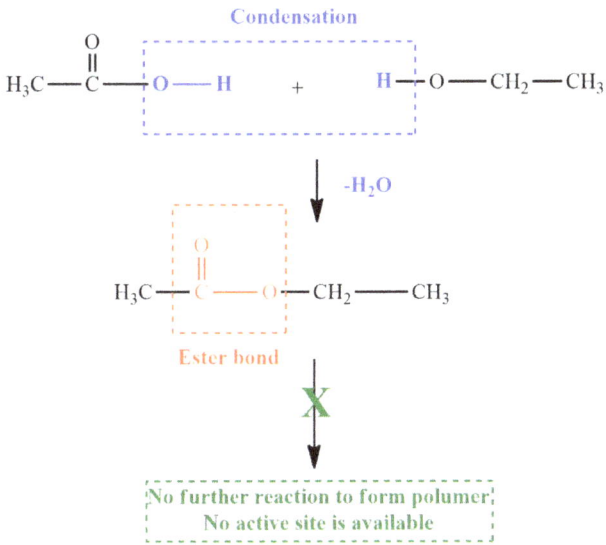

Figure 6.1: Monomers with only one functional group.

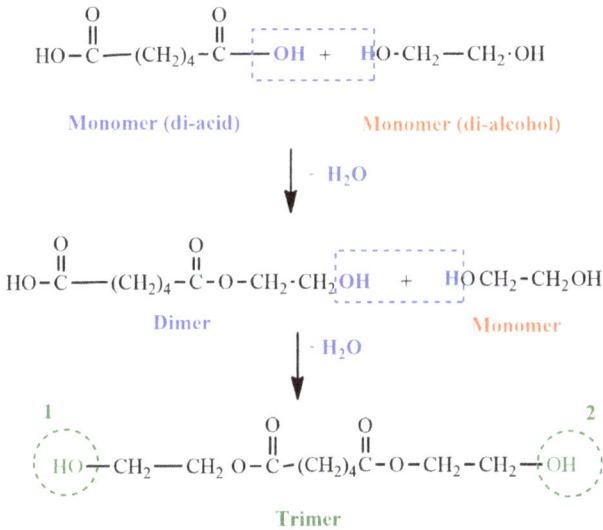

Figure 6.2: Bifunctional monomers.

6.3.1.1 Common Condensation Polymerizations

Polyesters: When dicarboxylic acid and di-alcohol react, a water molecule is re-moved, and an ester linkage is formed; an example is the formation of polyethylene terephthalate polymer from the reaction of terephthalic acid with ethylene glycol (Figure 6.3).

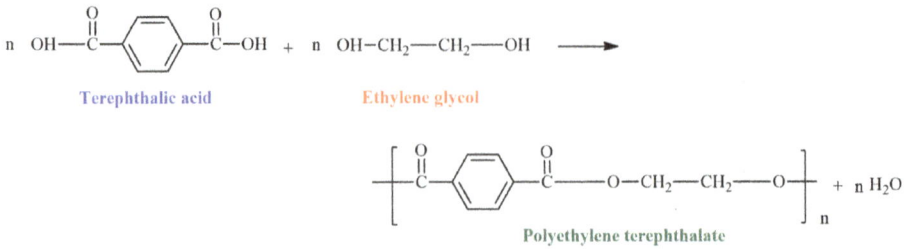

Figure 6.3: Formation of polyethylene terephthalate polymer.

Polyamides: When dicarboxylic acid and diamine react, a water molecule is removed, and an amide molecule is formed; an example is the reaction of adipic acid with 1,6-diaminohexane to form nylon-6, 6 (Figure 6.4).

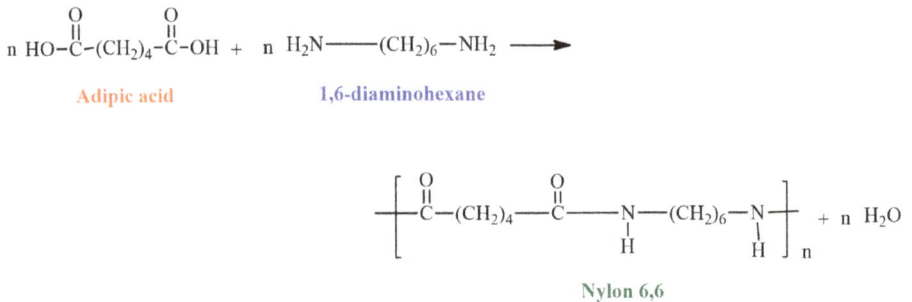

Figure 6.4: Formation of nylon-6, 6 polymer.

Polycarbonates: When bisphenol A and phosgene $COCl_2$ react, the HCl molecule is removed, and polycarbonate polymer is produced (Figure 6.5).

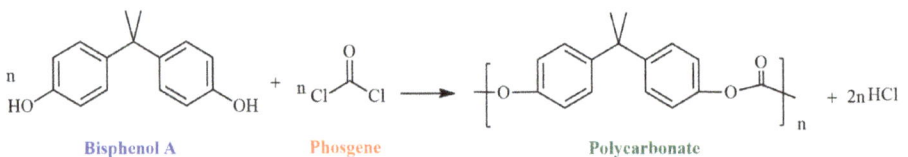

Figure 6.5: Formation of polycarbonate polymer.

6.3.2 Addition Polymerization

In this type of polymerization, the RU of the polymer contains the same atoms as the monomer. The majority of addition polymers are formed from monomers that at least have one double C=C bond.

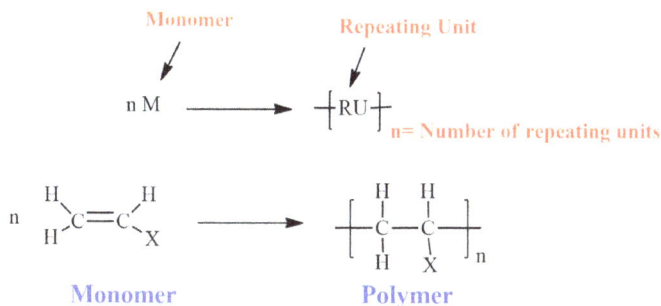

The addition polymerization reaction consists of three steps, namely:
- Initiation
- Propagation
- Termination

For addition polymerization, the initiation of chains may occur via free radical, cationic, anionic, or coordinate-acting initiators, but some monomers will be polymerized by more than one mechanism. Addition polymerizations are therefore classified as follows:
- Free radical polymerization
- Ionic polymerization (cationic and anionic)
- Coordination polymerization

6.3.2.1 Free Radical Polymerization

Free radical is the most commonly used method for the preparation of polymers from a wide variety of vinyl and diene monomers. The free radical polymerization is initiated by radicals (R·) and propagated by macro-radicals (RCH$_2$CH$_2$·) (Figure 6.6). These radicals exhibit an unpaired electron. Initiating radicals are thermally, electro-chemically, or photochemically formed from the added initiator. Initiation involves the generation of free radicals capable of reacting with monomers. Propagation is a pro-gressive addition of the monomers to the growing chain.

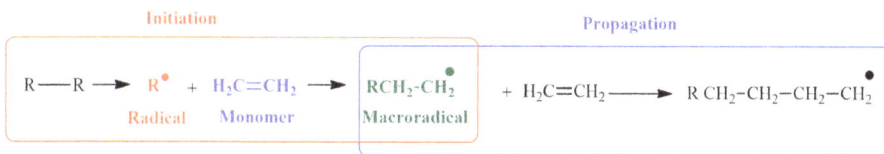

Figure 6.6: Free radical polymerization initiation and propagation steps.

Termination is a radical termination step (Figure 6.7), which can occur by the combi-nation (*two propagating radical chains meet and join at their free radical ends*) or by disproportionation (*two propagating radicals meet and undergo disproportionation,*

Chain transfer

$$•CH_2CH_3 + RCH_2CH{=}CH_2 \longrightarrow H_2C{-}•CH_2 +$$

Combination

$$•CH_2CH_2CH_2R + •CH_2CH_2CH_2R \longrightarrow RCH_2CH_2CH_2CH_2CH_2CH_2R$$

$$RCH_2C{-}•CH_2 + \quad H$$

Disproportionation

$$RCH_2CH{=}CH_2 + CH_3CH_2CH_2R$$

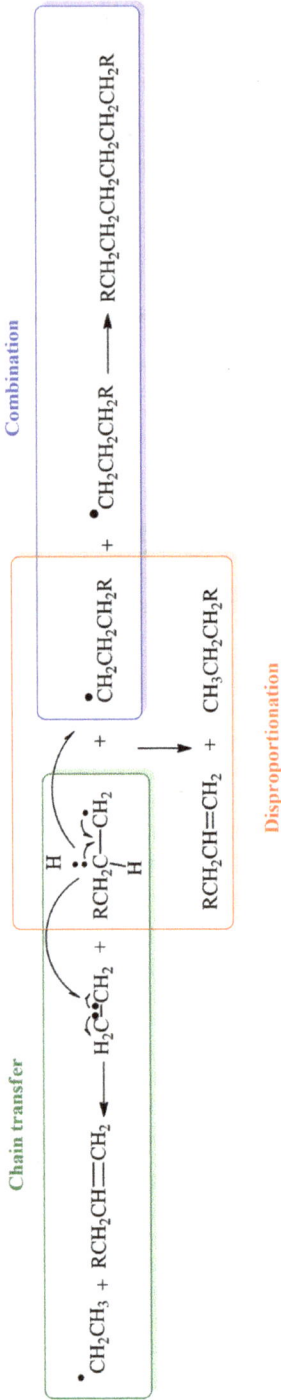

Figure 6.7: Free radical polymerization termination step.

one being oxidized to alkene, and one being reduced to alkane), or by chain transfer reaction (*free radical electron abstracts single electron from the growing polymer chain*).

6.3.2.2 Ionic Polymerization (Cationic and Anionic)

Cationic polymerization is a chain-growth polymerization in which a monomer is activated through a transfer of charge by a cationic initiator. This reactive monomer reacts further with other monomers to form a polymer (Figure 6.8). The stability of the carbocation is crucial in both initiating and propagating steps. Examples of effective catalysts are $AlCl_3$, BF_3, $TiCl_4$, and $SnCl_4$.

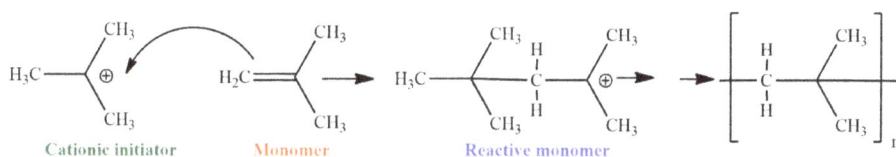

Figure 6.8: Cationic polymerization.

In an anionic polymerization, an anionic initiator transfers an electron (or negatively charged group) to a monomer which then becomes reactive (Figure 6.9). Many vinyl monomers with electron-withdrawing groups readily undergo anionic polymerization.

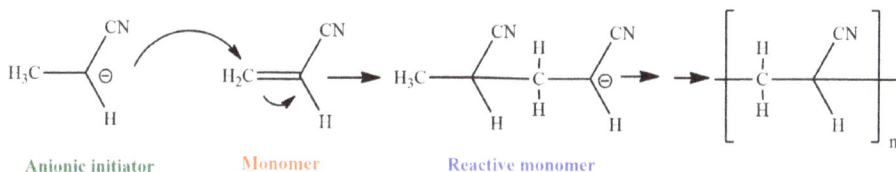

Figure 6.9: Anionic polymerization.

6.3.2.3 Coordination Polymerization

In the coordination polymerization, the kinetic chain starts with the addition of a monomer to a metal complex, propagation is by successive insertion of monomer at the metal, and termination occurs when the metal complex separates from the polymer.

For example, high-density PE (HDPE) is produced by coordination polymerization using a catalyst, usually Ziegler–Natta (ZN).

It can be prepared by a number of routes:
- By the coordination of metal complexes through functional groups
- By reaction with polymer-containing ligands
- By polymer formation through chelation

6.4 Copolymerization

A copolymer is made by copolymerization from more than one type of monomer that has similar or very different physical properties. This will result in a large variety of copolymers with very different properties and end-uses. Copolymers can be classified according to their monomer (M1/M2) sequence (Figure 6.10).

Randon   M1M1M2M2M1M2M1M1M2

Alternating   M1M2M1M2M1M2M1M2M1

Block   M1M1M1M2M2M2M1M1M1

Graft   M1M1M1M2M2M2M1M1M1
                M1   M1   M2   M1   M1
                M1   M1   M2   M1   M1
                M1   M1   M2   M1   M1
                M1   M1   M2   M1   M1

Figure 6.10: Copolymers classification.

6.5 Degree of Polymerization

It is defined as the number of RUs linked together by chemical bonds. The **molecular weight** (MW) of any given polymer can be calculated from the MW of the RU and the degree of polymerization (DP).

In a Polyvinyl Chloride (PVC) structure, If n = DP =1000 and MW of RU =
$$(2xC) + (3xH) + Cl = 2x12 + 3 + 35.5 = 24 + 38.5 = 62.5$$
$$DP = MW \text{ of polymer/ MW of RU}$$
$$1000 = MW \text{ of polymer } / 62.5$$
Molecular weight (MW) of polymer = MW of RU x DP = 62.5 x 1000 = 62,500

Repeating unit derived from $CH_2=CHCl$ monomer

Degree of polymerization

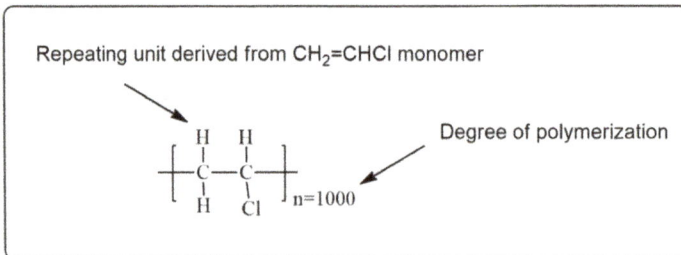

6.6 Glass Transition Temperature

Physical properties of polymers undergo large changes at important temperature, namely, the glass transition temperature (T_g) "*the temperature at which brittle material changes to a soft and flexible one.*" Examples of certain polymers and their T_g temperatures are listed in Table 6.1.

Table 6.1: T_g of certain polymers and their abbreviations.

Acronym	Name	T_g (°C)
PE	Polyethylene	−125
PP	Polypropylene	−10
PS	Polystyrene	100
PVC	Polyvinyl chloride	81
NR	Natural rubber	−75
PC	Polycarbonate	150

6.7 Polymerization Techniques

There are four polymerization techniques.

Bulk: This is carried out by adding a soluble initiator to pure monomer in the liquid state. This technique is used for the production of PS, LDPE, and poly(methyl methacrylate).

Solution: Monomer and initiator must be soluble in the liquid and the solvent must have the desired chain transfer characteristics and boiling points. PVC and polybutadiene are prepared by this technique.

Suspension: A water-insoluble monomer and an initiator are used. Polymers produced by this technique include PVC and copolymers such as poly(styrene–co–acrylonitrile) and poly(vinyl chloride–covinylidene chloride).

Emulsion: In this technique, the system usually contains a water-soluble initiator, chain transfer agent, and surfactant. Styrene–butadiene rubber and acrylonitrile–butadiene–styrene terpolymer are usually produced by this technique.

6.8 Stereochemistry of Polymers (Tacticity)

6.8.1 Isotactic: Substituents or side groups (chloride atoms or phenyl groups) are on the same side of the backbone (Figure 6.11).

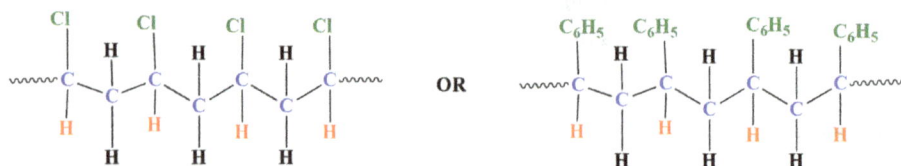

Figure 6.11: Isotactic polymers.

6.8.2 Syndiotactic: Substituents or side groups (chloride atoms or phenyl groups) are on the alternating sides of the backbone (Figure 6.12).

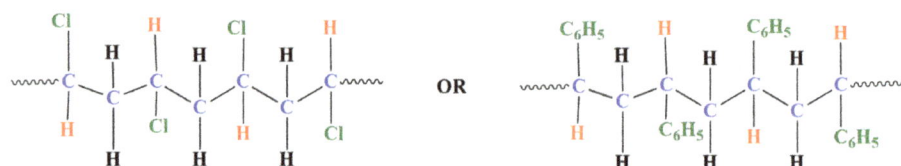

Figure 6.12: Syndiotactic polymers.

6.8.3 Atactic: Substituents or side groups (chloride atoms or phenyl groups) are on random sides of the backbone (Figure 6.13).

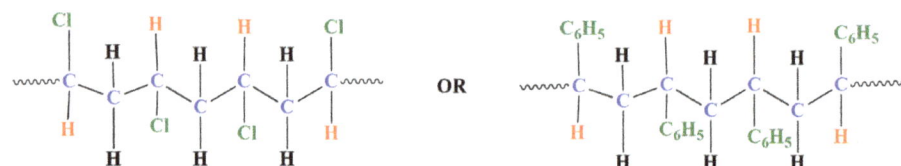

Figure 6.13: Atactic polymers.

6.9 Vulcanization

Vulcanization is a chemical process in which elastomers such as natural rubber (NR) are cross-linked by heating with sulfur (Figure 6.14).

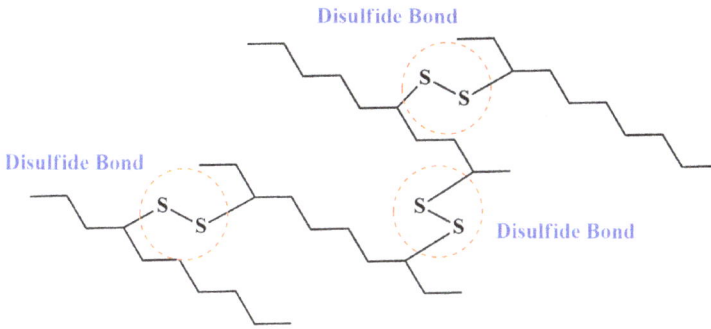

Figure 6.14: Vulcanization.

6.10 Commonly used Abbreviations for Selected Industrial Polymers

Abbreviations	Common names
ASA	Acrylonitrile–styrene–acrylate terpolymer
CPVC	Chlorinated polyvinyl chloride
PAN	Polyacrylonitrile
PET/PETE	Polyethylene terephthalate
PVA	Polyvinyl acetate
PVB	Polyvinyl butyral
SAN	Styrene–acrylonitrile copolymer
PVC	Polyvinyl chloride
PVF	Polyvinyl fluoride
PU/PUR	Polyurethane
PTFE	Polytetrafluoroethylene
PPG	Polypropylene glycol
PE	Polyethylene
PS	Polystyrene
PC	Polycarbonate

6.11 Polymer Structure and Properties

The physical properties of polymers can be explained using the same concepts encountered with small molecules. Although polymers do not crystallize or melt quite the same way smaller molecules do, crystalline regions in a polymer can be detected, and the temperature at which these *crystallites* melt can be measured.

6.11.1 Polymer Crystallinity

Large crystals are rarely formed by polymers, but many form microscopic crystalline regions called crystallites. A highly crystalline polymers are generally denser, stronger, and more rigid than similar polymers with a lower degree of crystallinity. PE shows a good example of how crystallinity affects a polymer's physical properties, for example, the highly branched, LDPE forms very small crystallites but the unbranched, HDPE forms larger and stronger crystallites. Because the HDPE has a higher degree of crystallinity, it is, therefore, denser, stronger, and more rigid than LDPE.

6.11.2 Thermal Properties

Long-chain polymers are glasses at low temperatures, and they go through a glass transition temperature, abbreviated T_g once the temperature is raised. Above the glass transition temperature, a highly crystalline polymer becomes flexible and moldable. When the temperature is raised further, the polymer reaches the crystalline melting temperature, abbreviated T_m. At this temperature, crystallites melt, and the individual molecules can slide past one another. Above T_m, the polymer is a viscous liquid and can be extruded through spinnerets to form fibers.

6.11.3 Plasticizers

When the polymer is too brittle because its glass transition temperature is above room temperature or because it is too highly crystalline, the addition of a plasticizer is a nonvolatile liquid that dissolves in the polymer, lowering the attractions between the polymer chains and allowing them to slide by one another. It acts by reducing the crystallinity of the polymer and lowering its glass transition temperature.

PVC is an example of a plasticized polymer. The common atactic form has a T_g of about 80 °C. Dibutyl phthalate is added to the polymer to lower its glass transition temperature to about 0 °C and makes it flexible. Once the soft, plasticized vinyl gradually loses its plasticizer, it becomes hard and brittle.

6.12 Biodegradable Polymers

The high chemical stability of many polymers is both a blessing and a curse. Heat resistance, wear resistance, and long life are valuable characteristics of clothing fibers, bicycle helmets, underground pipes, food wrappers, and many other items. Yet when those items outlive their usefulness, their disposal becomes a problem. The best solution is to recycle unwanted polymers. For example, soft-drink bottles are made from recycled poly(ethylene terephthalate), trash bags are made from recycled LDPE, and garden furniture is made from recycled PP and mixed plastics. Because many plastics are simply thrown away rather than recycled, developing biodegradable polymers that can be broken down rapidly by soil microorganisms will solve the problem. Among the most common biodegradable polymers are polyesters such as polyglycolic acid, polylactic acid, and polyhydroxy butyrate. They have found a wide range of uses.

6.13 Applications of Industrial Polymers

Depending on various applications, most of the polymeric materials can be classified into five major groups, namely, plastics, fibers, elastomers (rubber), coatings, and adhesives.

6.13.1 Plastics

- Commodity plastics: LDPE, HDPE, PP, PVC, PS.
- Engineering plastics: acetal, polyamide, polyamideimide, polyacrylate, polybenzimidazole.
- Thermosetting plastics: phenol-formaldehyde, urea-formaldehyde, unsaturated polyester, epoxy, melamine-formaldehyde.
- Functional plastics: optics, biomaterial.

6.13.2 Fibers

- Cellulosic: acetate rayon, viscose rayon.
- Noncellulosic: polyester, nylon (nylon-6, 6, nylon-6).
- Olefin: PP, copolymer (PVC 85% + PAN and others 15%; vinyon).
- Acrylic: contain at least 80% acrylonitrile (PAN 80% + PVC and others 20%).

6.13.3 Rubber (Elastomers)

- NR: *cis*-polyisoprene.
- Synthetic rubber: styrene–butadiene, polybutadiene, ethylene–propylene, poly-chloroprene, polyisoprene, polyurethane.
- Thermoplastic elastomer: styrene–butadiene blocks copolymer.

6.13.4 Coating and Adhesives

- Coating: lacquer, vanishes, paint, latex.
- Adhesives: solvent based, hot melt, pressure sensitive such as acrylate, epoxy, urethane, cyanoacrylate.

6.14 Polymer Additives

- Antioxidants: decompose hydroperoxides to prevent thermal degradation of polymers. Examples are hydroperoxide decomposers and hindered amine stabilizers.
- Light stabilizer: ultraviolet stabilizer (and hindered amine light stabilizers). Examples are 2-hydroxybenzophenones and 2-hydroxyphenylbenzo-triazoles.
- Nucleating agents: nucleators improve mechanical properties such as stiffness heat distortion temperature and enhance crystallization rate. Examples are di-benzylidene sorbitol (DBS) and sodium benzoate.
- Metal deactivators: prevent oxidative degradation that is caused by copper catalyst. Examples are citric acid and ethylenediaminetetraacetic acid
- Plasticizers: they improve the flexibility of polymer material. Examples are dioctyl phthalate and diisononyl adipates.
- Lubricants: they generally prevent the polymer from sticking to the processing equipment. Examples are PP waxes, PE waxes, and distearylphthalate.
- Acid scavengers: improve polymers sustainability and shape them to resist acids and radicals. Examples are calcium stearates and calcium lactates.
- Antiblocking additives: reduce blocking at the surface of polymer films and allow easier and smooth processing and handling. Examples are synthetic silica gel, limestone, and zeolites.
- Slip additives: usually used to reduce a film's resistance to sliding over it or parts of converting equipment. Examples are erucamides and oleamides.
- Antifogging additives: antifog additives are used to prevent fogging in plastic films, especially in food packaging and agricultural (greenhouse film) applications. Examples are ethoxylated sorbitan ester and ethoxylated alcohol.

– Antistatic additives: Antistatic additives are added to plastics to reduce or eliminate a static buildup. Antistatic additives work by lowering the resistivity of material so that charges are mobile and therefore will not cause static issues such as dust attraction or electrostatic discharge. Examples are ethoxylated amine and glycerol monostearate.

6.15 Summary and Important Concepts

Polymer. A polymer is a substance built up of a number of repeating chemical units held together by chemical bonds.

Addition polymer. This is a polymer that results from the addition of monomers to the growing chain (usually unsaturated molecules with double or triple bonds).

Condensation polymer. This is a polymer that is formed by the combination of two bifunctional monomers and releasing of small molecules.

Chain reaction. This is a reaction that proceeds by stepwise mechanism, in which each step generates macroradical that causes the next step to occur.

Linear polymers. These are polymers similar in structure and form long straight chains of identical links connected together.

Branched chain polymers. These polymers have branches originating at random points from the single linear chain.

Cross-linked or network polymers. These are polymers that have monomers linked together to form a 3D structure or network.

Elastomers. Elastomers are rubber-like solid polymers that are elastic in nature.

Thermoplastic polymers. These are long-chain polymers which when heated are softened (thick fluid-like) and hardened when cooled down.

Fibers. Fibers are thread-like in nature and can easily be woven.

Thermosetting polymers. These are semi-fluid in nature with low molecular masses. They form 3D structures on the application of heat.

Polymerization. This is a process by which small molecules (monomers) are converted to large molecules (polymers).

Cationic polymerization. This is a chain-growth polymerization in which a monomer is activated through a transfer of charge by a cationic initiator.

An anionic polymerization. This polymerization proceeds when an anionic initiator transfers an electron (or negatively charged group) to a monomer to become reactive.

Copolymer. This is a polymer made by copolymerization from more than one type of monomer that has similar or very different physical properties.

Degree of polymerization. This is the number of RUs linked together by chemical bonds.

Glass transition temperature. This is the temperature at which brittle material changes to a soft and flexible one.

Bulk polymerization. This is a technique that is carried out by adding a soluble initiator to pure monomer in the liquid state.

Solution polymerization. This is a technique in which both monomer and initiator are soluble in the liquid and the solvent have the desired chain transfer characteristics and boiling points.

Suspension polymerization. This is a technique in which a water-insoluble monomer and an initiator are used.

Emulsion polymerization. In this technique, the system usually contains a water-soluble initiator, a chain transfer agent, and a surfactant.

Isotactic polymer. This is a polymer that has substituents or side groups on the same side of the backbone.

Syndiotactic polymer. This is a polymer that has substituents or side groups on the alternating sides of the backbone.

Atactic polymer. This is a polymer that has substituents or side groups on random sides of the backbone.

Vulcanization. This is a chemical process in which elastomers such as NR cross-link by heating with sulfur.

6.16 Practice Exercises

6.16.1 What are the two main characteristics of the condensation polymerization monomers?

6.16.2 List down the main steps in the free radical polymerization.

6.16.3 Describe the stereochemistry of the atactic polymers.

6.16.4 Give three examples for polymer additives.

6.16.5 Classify industrial polymers on the basis of molecular forces.

6.16.6 What are the main differences between isotactic and syndiotactic polymers?

6.16.7 Which of the following is the best initiator for an anionic polymerization?
i.   BuLi
ii. PhOH
iii. BF_3

6.16.8 Is the polymer shown below an addition polymer or a condensation polymer?

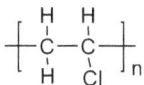

6.16.9 Which of the following monomers is most well-suited to cationic polymerization?
i.   Isobutylene
ii. Ethylene
iii. Acrylonitrile

6.16.10 Draw the structure of poly(acrylonitrile).

6.16.11 What is the chemical structure of Teflon?

6.16.12 Which of the following is least readily formed by cationic polymerization?
i.   Poly(acrylonitrile)
ii. Poly(isobutylene)
iii. PS

6.16.13 If the side groups of a polymer chain are generally on the same side of the polymer backbone, the polymer is called:
i.   atactic.
ii. syndiotactic.
iii. isotactic.

6.16.14 Draw the chemical structure of isotactic PP.

6.16.15 Draw the chemical structure of syndiotactic PS.

6.16.16 Which of the following elements is necessary to the vulcanization of rubber?
i.   Sulfur
ii. Phosphorus
iii. Boron

6.16.17 Which of the following is a condensation polymer?
i.   Poly(ethylene terephthalate)
ii. Poly(tetrafluoroethylene)
iii. PS

6.16.18 A substance that is added to a polymer to lower its T_g is called a _____.
i.   plasticizer
ii. thermolite
iii. copolymer

6.16.19 What kind of stereochemistry do the following polymers have?

OR

6.16.20 Write the full names of the following polymers:
i.   PAN
ii. SAN
iii. PPG

6.16.21 Give two examples of light stabilizers.

6.16.22 What kind of initiator is used in the emulsion technique?

6.16.23 Calculate the MW of the PVC polymer in which the DP is equal 1000.

6.16.24 Show the main differences between random and alternating copolymers.

6.16.25 Give two examples for the catalysts that are used in cationic polymerization.

Abbreviations

BOC	Butyloxycarbonyl
BPA	Bisphenol A
CA	Cellulose acetate
CAP	Cellulose acetate phthalate
CMC	Carboxymethylcellulose
CN	Cellulose nitrate
CODIS	Combined DNA Index System
CPG	Controlled pore glass
CS	Cellulose sulfate
DBS	Dibenzylidene sorbitol
DCC	Dicyclohexylcarbodiimide
DIC	Diisopropylcarbodiimide
DNA	Deoxyribonucleic acid
dNTPs	Deoxynucleotide triphosphates
DP	Degree of polymerization
EC	Ethyl cellulose
ESD	Electrostatic discharge
GNA	Glycol/glycerol nucleic acids
HDPE	High-density polyethylene
HEC	Hydroxyethyl cellulose
HPC	Hydroxypropyl cellulose
HPMC	Hydroxypropyl methylcellulose
LCPC	Low crystallinity powdered cellulose
LDPE	Low-density polyethylene
MCC	Microcrystalline cellulose
MC	Methylcellulose
MNA	Morpholinos nucleic acid
mRNA	Messenger RNA
NR	Natural rubber
PA	Polyamide
PAA	Polyacrylic acid
PC	Polycarbonate
PC	Powdered cellulose
PCR	Polymerase chain reaction
PE	Polyethylene
PEO	Polyethylene oxide
PET	Polyethylene terephthalate
pI or IEP	Isoelectric point
PLA	Polylactic acid
PNA	Peptide (or peptido) nucleic acid
PP	Polypropylene
PPO	Polypropylene oxide
PS	Polystyrene
PVA	Polyvinyl acetate
PVC	Polyvinyl chloride
RNA	Ribonucleic acid
RU	Repeating units

https://doi.org/10.1515/9783110762761-007

SBR	Styrene–butadiene rubber
4'-SDNA	4'-Thionucleic acids
SS	Solid support
STR	Short tandem repeat
T_g	Glass transition temperature
T_m	Crystalline melting temperature
TNA	Threose nucleic acid
ZN	Ziegler–Natta

Resources and Further Readings

[1]   Wade LG. Organic Chemistry, Pearson, 2014, ISBN: 9781292021652.
[2]   McMurry J. Organic Chemistry, Brooks Cole, 2011, ISBN: 9780840054449.
[3]   Stoker HS. General, Organic, and Biological Chemistry, Brooks Cole, 2010,
      ISBN: 9780618606061.
[4]   McMurry J, Castellion ME, Ballantine DS, Hoeger CA, Virginia E. Fundamentals of General,
      Organic, and Biological Chemistry, Peterson, 2007, ISBN: 0136054501.
[5]   Morrison RT, Boyed RN. Organic Chemistry, Prentice Hall, Englewood Cliffs, NJ, 1992,
      ISBN: 0136436692.
[6]   Solomons TWG, Fryhle CB, Snyder SA. Organic Chemistry, John Wiley & Sons, 2016,
      ISBN: 9781118875766.
[7]   Zweifel H. Plastics Additives Handbook, Hanser Publications, 2001, ISBN: 3446216545.
[8]   Carraher CE Jr. Seymour Carraher's Polymer Chemistry, CRC Press, Taylor & Francis Group,
      2006, ISBN: 9781420051025.
[9]   Blackburn GM, Gait MJ, Loakes D, Williams DM. 2006, Nucleic Acids in Chemistry
      and Biology, RSC Publishing, 2006, ISBN: 9780854046546.
[10] Lönnberg H. Chemistry of Nucleic Acids, De Gruyter, 2020, ISBN: 9783110609271.

https://doi.org/10.1515/9783110762761-008

Index

https://doi.org/10.1515/9783110762761-009

www.ingramcontent.com/pod-product-compliance
Lightning Source LLC
Chambersburg PA
CBHW081545220326
41598CB00036B/6573